Dedication

To my two sons, Samuel and Benjamin, and their mother, Jan.

From Regulation to Competition

From Regulation to Competition: New frontiers in electricity markets

edited by
Michael A. Einhorn
U. S. Department of Justice

Kluwer Academic Publishers
Boston / Dordrecht / London

Distributors for North America:
Kluwer Academic Publishers
101 Philip Drive
Assinippi Park
Norwell, Massachusetts 02061 USA

Distributors for all other countries:
Kluwer Academic Publishers Group
Distribution Centre
Post Office Box 322
3300 AH Dordrecht, THE NETHERLANDS

Library of Congress Cataloging-in-Publication Data

From regulation to competition: new frontiers in electricity markets
    / edited by Michael A. Einhorn.
        p. cm.
    Includes bibliographical references and index.
    ISBN 0-7923-9456-9
    1. Electric utilities—Government policy—Great Britain.
2. Electric utilities—Government policy—United States.
3. Privatization.   I. Einhorn, Michael A.
HD9685.G72F74   1994
333.79'32'0941—dc20   94–9265
                                                                    CIP

Printed on acid-free paper.

Printed in the United States of America

Contents

Contributing Authors

BARKER, James V., Jr., Vice President and Executive Consultant, Management Consulting Services, ECC, Inc., Fairfax, Virginia USA

BRAMAN, Susan P., Economist, Bureau of Economics, Federal Trade Commission, Washington, D.C., USA

DUNN, William H., Jr., Principal Consultant, Management Consulting Services, ECC, Inc., Fairfax, Virginia, USA

EINHORN, Michael A., Economist, Antitrust Division, U.S. Department of Justice, Washington, D.C., USA

GREEN, Richard J., Junior Research Officer, Department of Applied Economics, Fitzwilliam College, Cambridge, England

HOGAN, William W., Thornton Bradshaw Professor of Public Policy and Management, John F. Kennedy School of Government, Cambridge, Massachusetts 02139, USA

HUNT, Sally S., Director, National Economic Research Associates, Inc., London, England

KAHN Edward P., Leader, Utility Policy and Planning Group Lawrence Berkeley Laboratory, Berkeley, California, USA

LITTLECHILD, Stephen C., Director General of Electricity Supply Office of Electricity Regulation, Birmingham, England

PEREZ-ARRIAGA, Ignacio J., Professor of Electrical Engineering Instituto de Investigacion Tecnologica, Universidad Pontificia Comillas, Madrid, Spain

RUFF, Larry E., Managing Director, Putnam, Hayes, and Bartlett, Inc., Washington, D.C., USA

SHEAN, Mk P., International Project Services, Inc., 3220 N. Street N.W., Suite 173, Washington, D.C., USA

SHUTTLEWORTH, Graham, Associate Director, National Economic Research Associates, Inc., London, England

VICKERS, John S., Drummond Professor of Political Economy, Nuffield College, Oxford, England

VOGELSANG, Ingo, Professor of Economics, Boston University, Boston, Massachusetts, USA

YARROW, George K., Director, Regulatory Policy Research Center Hertford College, Oxford, England

1 INTRODUCTION

Michael A. Einhorn[1]

1.1. Introduction

Electric utilities throughout the world continue to face new challenges involving ownership, market structure, and regulation. The papers included in this volume bear directly on three related issues: (1) Should ownership be private or public? (2) What operations should be integrated, and where is competition feasible? (3) Where is regulation necessary, and can it be made more efficient?

Chapters 2 through 7 discuss the British electricity experiment that has privatized and disintegrated the nation's generation, transmission, and distribution companies, introduced market competition for power purchases, and implemented incentive regulation for monopolized transmission and distribution grids. The remaining chapters focus on the theater in which significant microeconomic issues will continue to emerge, most immediately in the United Kingdom and United States, regarding the coordination and pricing of transmission.

1.2. The British Electricity Experiment

Britain's Electricity Act of 1947 established the Central Electricity Authority as a public corporation to manage the country's integrated generation and transmission companies; fourteen independent area boards were created to manage separate regional distribution grids. The Electricity Act of 1957 established the Central Electricity Generating Board (CEGB) as an integrated generation and supply entity and the Electricity Council as a regulatory watchdog body that facilitated coordination and set rates. Bulk supply tariffs (BSTs) for distributor purchases of power supply in the vertically nonintegrated system were based on long-run marginal cost.

The industry implemented two significant reforms in the early Thatcher years. First, the Electricity Council adopted the recommendations of a 1978 White Paper to shift the emphasis of policy objectives from pricing and investment to increased profitability and cash flow; this provided greater incentives for company managers to improve financial performance. Second, the Energy Act of 1983 required each area board to purchase power from private producers other than the CEGB at rates based on its avoided cost through private purchase tariffs (PPTs). Transmission and distribution networks were also partially opened for wheeling transactions.

The effect of the two reforms was disappointing. First, while financial performances improved, generation capacity was excessive, cost overruns were large, and annual improvement rates in total factor productivity did not change from magnitudes over the previous twenty years (Vickers and Yarrow, 1988, p. 287). Second, when the Energy Act was enacted in 1983, avoided cost tariffs could have been based on prevailing BST rates that included long-run capacity and fuel costs. However, revenue requirements for embedded generation capacity instead were reassigned to area boards for recovery; assignments were partially based on relative maximum demands in an historic test year—1982. Based on historic-test-year demands, these cost assignments were subsequently nonavoidable; this cost assignment consequently reduced measured avoided costs and the resulting incentives for independent suppliers.[2]

In this context, the British government issued another White Paper in 1988 that prescribed new objectives for restructuring the nationalized CEGB and area boards. Two primary principles were established: economic efficiency is the principal objective of the reform strategy, and competition is the best way to achieve this goal. Based on the 1988 White Paper, the Electricity Act became law in July 1989 and instituted industry restructure in March 1990.

Under the industry restructure, the twelve public area boards in England and Wales became twelve private regional electricity companies (RECs). The CEGB was broken into three generation companies—National Power (capacity share 52 percent, primarily fossil fuel), PowerGen (33 percent, fossil fuel), and Nuclear Electric (15 percent, nuclear)—and a centralized transmission grid (National Grid Company) that owns some pumped storage capacity; each REC may self-generate up to 15 percent of its power needs and may contract directly with any independent generator of less than 100 megawatts (MW). The government sold off 60 percent of PowerGen and National Power in 1991, but Nuclear Electric remained public; National Grid was turned over to the RECs but must be operated at arm's length. National Grid may purchase power from any independent generator, including new entrants.

In the integrated system, the CEGB had dispatched generation plants in each hour based on short-run marginal supply costs, which included generation and transmission components. The new system replaced this centralized dispatch with a spot market for wholesale electricity where competing generation companies bid to sell power to National Grid. At each day's end, each generator informs National Grid of its power supply prices for each half hour of the next day; the grid company then purchases power from low bidders. Each bidding generator is paid a pool input price (PIP) that comprises its bid system marginal price (SMP, presumably reflecting fuel and operating costs) for *actual power* supplied plus an add-on rider for all capacity *declared available*; the capacity element is the loss of load probability multiplied by the estimated net cost of a supply shortage,[3] and it provides incentives for generators to add capacity.

Buyers pay a pool output price (POP), which includes PIP plus instantaneous add-ons (called uplifts) to reflect the costs of transmission constraints, spinning reserves, and so on. Because pool input prices are volatile, generators and distributors now enter into forward contracts with specified strike prices to purchase power in advance of use; these "contracts for differences" now cover 95 percent of all power purchases. Finally, in order to protect Nuclear Electric, each REC must purchase a designated fraction of its demand from nonfossil generation sources.

The country's transmission and distribution sectors remained monopolized and were regulated by the newly created Office of Electricity Regulation (OFFER). Transmission prices depend exclusively on where the power is produced and consumed, and no allowance is made for different capital costs that new and existing generators may impose. Initial prices were based on embedded cost. In any year, National Grid may increase its average charges per kilowatt of transmission capacity by the percentage change in the retail price index (RPI), less a specified offset, X, to represent prospective improvements in company productivity.[4]

Each regional company was granted a monopoly to supply power to small customers in its franchise territory.[5] Each distributor was also granted rights to supply power to any large customer anywhere and was obliged to provide wheeling capacity to facilitate nonfranchised power supply; supply prices for these large customers were deregulated.[6] In each REC, overall distribution charges were price-capped at RPI + X; X allowances ranged from 0 percent to 2.5 percent. Supplementary price caps were enacted for power sold to franchised small customers (see Chapter 6 for details).

Since privatization, contracts have been signed for 22,000 MW of new generation capacity, mostly low-risk combined-cycle gas turbine; three coal plants have been cancelled. The move toward combined cycle is controversial; the North Sea has limited supplies of natural gas. In addition, the country now has a 60 percent reserve margin in generation capacity, contended by some to result from transmission prices that are too low and wheeling requirements that are too liberal.

1.3. Discussion of the British Experiment

Chapters 2 through 7 in this volume discuss the British experiment and possible future avenues of reform. Larry Ruff identifies three principal problems in the present structure: generation pricing rules can easily be gamed by a small group of producers; ex ante pool prices do not reflect efficient capacity charges; and transmission and generation dispatch are not properly integrated and unbundled transmission prices are inefficient. As a remedy, Ruff suggests that spot-market prices be based on the incremental cost of the marginal kilowatt hour that is actually supplied—that is, these prices should be determined ex post. He also argues that grid management be placed under the control of the producers and distributors who actually use it and pay its costs; if managed on behalf of these controlling principals, this controlling entity would have the correct incentives to dispatch and expand transmission plant efficiently.

John Vickers and George Yarrow consider some important practical problems from recent theoretical literature. First, when generation and transmission/distribution controllers hold up contracting in order to expropriate rents from nonrecoverable investments, generation owners may underinvest in necessary plant. Second, because generation sellers must precommit to capacity levels before establishing bid prices, competition in an oligopolistic market may be limited, and prices will not fall to marginal cost (as in Kreps and Scheinkman, 1983). Vickers and Yarrow identify three strategies for reversing these inefficient incentives—more use of long-term contracting, additional entry of suppliers, and regulatory intervention to protect first-movers and otherwise limit opportunistic behavior.

Developing a body of research on competitive generation bidding (Bolle, 1992; Green and Newbery, 1992; Helm and Powell, 1992; Powell, 1992; von der Fehr and Harbord, 1992a, 1992b), Richard Green examines the equilibrium outcome in greater detail. In a duopoly model with strategic interaction, Klemperer and Mayer (1989) show that each producer should set bid price above (below) marginal cost if spot-market demand is greater (less) than constrained capacity. Green examines data that confirm this theoretical insight. As a result, Green contends that bid prices could be too high and dispatch could be inefficient in the short run. Furthermore, the industry might not be sufficiently competitive to reverse these inefficiencies in the long run. These fears have recently been heeded; the two largest generating companies agreed in late 1993 to sell off 6,000 MW of existing capacity, thereby doubling the amount of independent power on the system. These companies also agreed to bid into the pool in a manner that could reduce average purchase prices up to seven percent.

Sally Hunt and Graham Shuttleworth consider the National Grid Company's tariff structure and regulation, which they contend is inefficient. Because transmission prices do not reflect incremental cost, NGC now earns no additional revenue to cover additional investments that must be incurred to expand capacity and operate efficiently. Because transmission and generation are nonintegrated, NGC fails to internalize the costs of line losses, which are passed back to generation companies and their customers. Wherever transmission prices are below incremental cost, generators have excess incentive to add more capacity; as a result, excess generation and transmission investment has resulted in the north. The authors suggest several strategies for reform, including spot-market marginal cost pricing and adjustments of allowed revenues based on demand growth.

Chapter 6 (written by myself), considers incentive regulation of Britain's twelve regional electric companies that were spun off in 1990 to provide distribution service to ultimate customers. Recognizing the importance of lumpy capital investments in electricity transport systems, the chapter contends that OFFER may be discouraging necessary and efficient investments by not adjusting prices to reflect estimated or actual capital costs. Faced with conflicting regulatory needs to share risks equitably and to maximize incentive, the chapter further contends that no benchmark of capital costs is reasonable and that pass-throughs of actual capital costs are a necessary evil. Because transport capital and power purchases are substitutable to some degree, this then requires a modification of OFFER's benchmark adjustment for purchase costs as well.

Director General Steven Littlechild of the Office of Electricity Regulation is appropriately given the last word on the British electricity experiment in order to respond to his critics. He points out that competition in generation is emerging at a reasonable pace; the combined market share for National Power and PowerGen has declined from 78 percent to 68 percent, and many new independent generating stations are now being built. Furthermore, the two large companies have moved from nuclear and coal construction to more efficient gas-fired combined cycle turbines; since Vesting in 1990, they have dramatically increased output productivity and halved their manpower needs. Bidding problems do remain; the pool price recently increased in April and May 1993. Littlechild contends that competitive supply to the 5,000 consumers with maximum demands above 1 MW is very successful; one-third of these customers (accounting for one-half of usage in the competitive market) have chosen electricity suppliers other than their local regional company. In all, real electricity prices have fallen since vesting, most dramatically for medium and moderately large customers.

1.4. Transmission Issues

The remaining chapters in this book deal with topics related to transmission systems and pricing, which are now key issues in the United States, Europe, Australia, and New Zealand. The U.S. Federal Energy Regulatory Commission (FERC) has required company filings of open-access tariffs as a precondition for approval of both mergers (*Utah Power and Light Co.*, 45 FERC 61, 95 (1988); *Northeast Utilities Service Co.*, 56 FERC 61, 269 (1992)) and market-based rates (*Entergy Services, Inc.*, 58 FERC 61, 234 (1992); *Pennsylvania Electric Co.*, 58 FERC 61, 278 (1992)). The 1992 Energy Policy Act gave to these regulators the authority to mandate wheeling for wholesale customers. Inefficient price setting in the British grid has been cited as a cause for overbuilding and mislocated generation capacity (see Chapter 4). Australia and New Zealand have spun off independent transmission grids and must now determine a means for ensuring efficient prices. Europe now aims to expedite international power trade between net exporters (such as France and Norway), net importers (Italy and Portugal), and distribution centers (Belgium and Switzerland) as part of the Unified European Market; as in the United States, contractors may need to recognize loop flow across third-party transmission grids.

Power transmission has several important attributes that must be considered in any pricing scheme:

1. Unlike natural gas, electricity is not conveniently stored; capacity requirements vary instantaneously with customer demand. Transmission rights can be acquired through long-term contracts or on a spot-market basis.

2. There are physical constraints on instantaneous real and reactive current loadings that limit allowed flow on any transmission circuit.

3. At any moment in time, system engineers attempt to meet customer loads by dispatching facilities in ascending order of short-run marginal cost. Economic dispatch is constrained by flow constraints, operating limits that disallow rapid ramp-up or ramp-down of generation plants, and safeguards that are necessary to ensure plant diversity needed for system reliability.

4. Power that flows from a generation source to any customer node moves across all available wires in inverse proportion to the resistance of each; loop flow between two interconnecting grids can lead to capacity shortages in either. When capacity shortages arise, the utility must reconfigure its dispatch order in a constrained and more costly manner.

5. Because electricity provision necessarily entails generation and transport costs, an efficient dispatch system should recognize coordination economies between the two sequential functions; this was an economic motivation for integration. Inefficient dispatch may arise under vertical separation unless prices reflect appropriate costs.

6. Spot-market and contract prices for transmission usage should allocate scarce capacity and should induce efficient system buildout and siting. These prices ideally should recognize customer location, capacity utilization of each affected wire, the overall configuration of customer loads on the transmission grid, loop flow externalities, and constraints on the authority to dispatch.

7. Electricity transmission systems exhibit significant economies of scale and are natural monopolies; marginal cost prices may inadequately recover revenues needed for capital recovery.

8. For a transitory period, regulators may need to incorporate price components to recover stranded investment costs that arise when customers forego utility generation for a wheeled-in alternative source. When utilities have obligations to serve (as in retail provision), additional standby prices may be necessary to allow for possible customer return.

James Barker, William Dunn, and Mark Shean consider the relevance of the British experiment to other nations and alternative paradigms for restructuring a country's electricity network. Electricity markets comprise six necessary operations—generation, transmission, distribution, system planning and operations, bulk power markets, and retail sales—that can be integrated in a number of different ways; illustrative structures are found in Great Britain, the United States, Northern Ireland, Holland, and Norway. Several important practical considerations should affect any policy maker's final choice: most important are permitted customer switches that may lead to stranded investments, asymmetrically informed and unbalanced negotiations between buyers and sellers, and the interrelationship and need for coordination between transmission and generation companies.

William Hogan's chapter takes a significant step toward resolving the appropriate pricing of electricity flow across a centrally controlled transmission network, such as Britain's; as discussed thoroughly by Ruff, the country's present strategy for transmission pricing is inefficient and dispatch responsibilities not properly integrated.[7] The matter is complicated by the physical properties of electricity, which flows simultaneously over all interconnecting network links and is not confined to one designated contract path; peak usage of a link can be constrained by thermal or voltage limits. Assuming efficient dispatch, Hogan shows that short-run marginal costs for transmission on a link are the difference between incurred and avoided generation costs at its two nodes; these short-run marginal costs are the efficient basis for instantaneous transmission prices. Based on these marginal costs, each owner of capacity rights on a particular link may collect revenues for use of its share by other parties. Assuming constant returns to scale, this pricing procedure also provides efficient incentives for capacity expansion.

Ignacio Pérez-Arriaga presents a theoretical regulatory framework of transmission access that addresses the seemingly conflicting issues of short-term economic efficiency, security of supply, autonomy of individual participants, and financial risk reduction; his results are meaningful in a single or multiple utility setting. His method combines the use of physically meaningful operation rules and a reduced set of prototype contracts for generation and transmission services. The chapter emphasizes large power systems that consist of interconnected entities that are independently dispatched with diverse ownership and coordination structures. His results may then be relevant to transmisson markets in America and the European continent where independent utilities provide power to consumers, interconnect with one another, and provide unwanted loop flow on each other's grid.

Susan Braman considers whether joint ownership of capacity rights in a natural monopoly transmission grid can be made competitive and improve allocative efficiency. In a competitive joint venture, anyone may become an owner of transmission rights in a new line by paying a capacity price based on unit average cost. Existing share owners of capacity along a particular line then may compete to provide power to wholesale or retail buyers. Braman demonstrates that this market solution achieves efficient capacity allocation in the short run and efficient scale in the long run.

As discussed by Edward Kahn, regional transmission groups (RTGs) may provide a cooperative way of addressing common concerns that arise under a regime of broader transmission access. The U.S. Energy Policy Act of 1992 (EPA) gave the Federal Energy Regulatory Commission the explicit authority under Section 211 of the Federal Power Act to order wheeling. It required transmission owners to file wheeling tariffs and to disclose relevant data regarding available capacity. Faced with transmission networks that interconnect independent control areas, the Commission now must determine strategies how to price transmission access and usage, reassign native load capacity, and compensate affected owners for nonavoidable loop flow. During the final stages of deliberation on the EPA, industry representatives presented consensus legislation that would have required the Commission to certify regional transmission groups whose primary purpose would be to facilitate the provision of wheeling and the resolving of disputes.[8] Kahn discusses institutional limits to nodal pricing advocated by Hogan and possible means of covering network expansion costs when scale economies exist.

Ingo Vogelsang considers some theoretical issues in the design of incentive mechanisms for independent transmission and distribution companies. He distinguishes basic and detailed regulatory engineering; the former category refers to the creation of a regulatory body that can execute detailed regulatory rules included in the latter. An inappropriate basic design may hinder the ability of government officials to make political commitments that may be needed to inspire confident investment. Vogelsang compares four detailed incentive mechanisms that may control overall utility prices: rate-of-return regulation, price-cap regulation, profit sharing, and yardstick regulation. He then discusses six ways to control individual prices within the utility's overall price basket.

Notes

1. Views are personal and not those of the U.S. Department of Justice.
2. By 1987, PPT prices were 90 percent of analogous BST prices (Vickers and Yarrow, 1988).

3. This is the difference between estimated value of lost load and SMP; the value of lost load is estimated at 2 pounds per kilowatt hour (kWh).

4. This is akin to the price-cap system for British Telecom that was implemented in 1984. The productivity offset was originally set at zero for the regional electric companies and 3 percent for British Telecom.

5. The initial threshold between small and large customers was 1 MW peak demand. This threshold will diminish to 100 kilowatts (kW) in 1994 and vanish entirely in 1998.

6. There are 4,500 customers with peak demands above 1 MW; these account for one-third of all electricity sales; 1,700 of these customers use a nonfranchised supplier.

7. Transmission pricing is now center stage in the United States as well; the Energy Act of 1992 gave to the Federal Energy Regulatory Commission (FERC) the explicit authority to mandate utility wheeling for wholesale transactions. (The Commission previously had declined to exercise an ambiguous prior mandate from the Public Utilities Regulatory Policy Act of 1978.) The Commission now must come up with a scheme for pricing usage and capacity rights.

8. Included among the criteria needed for certification are broad membership, coordinated planning and information sharing, fair procedures for dispute resolution, and obligations for capacity owners to wheel power and upgrade facilities (*Federal Register*, 58 (149): 41627).

References

Bolle, F. 1992. "Supply Function Equilibria and the Danger of Tacit Collusion: The Case of Spot Markets for Electricity." *Electricity Economics* (April): 94–102.

Green, R.J., and D.M. Newbery. 1992. "Competition in the British Electricity Spot Market." *Journal of Political Economy* 100(5): 929–953.

Helm, D., and A. Powell. 1992. "Pool Prices, Contracts, and Regulation in the British Electricity Supply Industry." *Fiscal Studies* 13(1): 89–105.

Klemperer, P. D., and M.A. Meyer. 1989. "Supply Function Equilibria in Oligopoly Under Uncertainty." *Econometrica* 57(6): 1243–1277.

Kreps, D., and J. Scheinkman, 1983. "Quantity Precommitment and Bertrand Competition Yield Cournot Outcomes." *Bell Journal of Economics* 14(2): 326–37.

Powell, A. 1992. "Trading Forward in an Imperfect Market: The Case of Electricity in Britain." Puper presented at the Royal Economic Society Conference, University College, London, April.

Vickers, J., and G. Yarrow. 1988. *Privatization*. Cambridge, Mass.: MIT Press.

Von der Fehr, N-H M., and D. Harbord. 1992a. "Long-Term Contracts and Imperfectly Competitive Spot Markets: A Study of the U.K. Electricity Industry." Memorandum No. 14. Department of Economics, University of Oslo.

———. 1992b. "Spot Market Competition in the U.K. Electricity Industry." Memorandum No. 9. Department of Economics, University of Oslo.

2 COMPETITIVE ELECTRICITY MARKETS
The Theory and Its Application
Larry E. Ruff

2.1. Introduction

Competition in electricity markets is a recent but rapidly growing phenomenon worldwide. It began to blossom in the United States as a largely unexpected result of the limited requirement that utilities buy power from independent power producers (IPPs) (actually, qualifying facilities, or QFs, in the U.S. legislation) at avoided cost, reached its fullest flowering to date in the privatization and competitive restructuring of the electricity supply sector in the United Kingdom, and is currently springing up in various forms elsewhere. Monopoly utilities around the world are battling the threatening weed whenever it appears on their home turf but are only delaying the day when it takes root in some evolved or mutated form.

This chapter discusses the theory and practice of competition in electricity markets, using the U.K. experience as an imperfect but instructive example. The competition at issue here is not limited to generators competing to sell to a monopoly utility with captive customers but includes allowing customers—wholesale or retail[1]—served by the wires of one utility to purchase their power from any of several competing suppliers. This form of competition requires fundamental changes in the organization and functioning of electricity systems and in the responsibilities of utilities and their customers; for example, a utility's traditional "obligation to supply" whatever its customers want must be redefined if those customers no longer have the "obligation to pay" whatever the utility spends. It is competition in supply—not just competition among IPPs to sell to a monopoly buyer or reseller—that promises the most benefits and that appears to be the ultimate objective of reformers in the United States and elsewhere. A good understanding of the theory and of its most complete real-world application in the United Kingdom—including an understanding of the design errors that have created problems there—is essential if mistakes are to be avoided elsewhere.

The chapter begins with a discussion of the fundamental conceptual problem in a competitive electricity market: how to maintain the central operational control that is essential on an electricity system for both technical and economic reasons, while allowing consumers and generators to deal commercially with each other on a bilateral basis. The theoretical solution to this problem is outlined briefly. Then the U.K. system is discussed, as evidence that even a radical and imperfect system based on this theory can work, at least in the sense that the lights stay on and investment continues. Finally, some ways to improve on and avoid the worst mistakes of the U.K. model are discussed.

2.2. Efficient Competition in Electricity

2.2.1. Coordination and Competition in Electricity Supply

The view that electricity supply is a natural monopoly was initially based on economies of scale in electricity generation: bigger machines have higher thermal efficiencies and lower unit costs. When electricity systems became large relative to the most economically efficient generating units, competition among generators was introduced, with dramatic and—despite some early mistakes—largely beneficial effects on the technology and business arrangements in the generation sector. The argument for natural monopoly then shifted to economies of scope and coordination: bigger systems, involving more diversified machines and consumers, can meet load reliably at lower cost but require nearly instantaneous control and coordination for technical and economic reasons. Thus, it was (and still is) argued that a single monopoly must own or purchase all generation, operate it to meet load at least cost, and resell a bundled product to customers who have no choice but to pay whatever costs the utility incurs in the process.

The argument that control by a monopoly is required to coordinate diverse buyers and sellers is not a priori persuasive. The whole point of markets is to coordinate diverse buyers and sellers, while bringing competitive forces to bear on them. True, the special nature of electricity requires specialized market arrangements that could prove so cumbersome and costly that leaving it to a monopoly may be the better solution. But it is certainly worth considering the possibility that modern theory and technology make it possible to design a competitive electricity market that would reduce some of the investment and pricing inefficiencies for which monopoly utilities are justly criticized. The economic issue is whether and how the special characteristics of electricity can be dealt with in a market that treats electricity much like any other commodity, with prices and commercial dealings between buyers and sellers coordinating most operational and investment matters.

2.2.2. The Need For a Dispatch-Based Spot Market

Any attempt to treat electricity as a commodity subject to standard economic and business principles must begin with two fundamental technical characteristics of electricity: (1) all the electricity in any single market must be moved and traded on a single, integrated grid; and (2) electricity cannot be stored, so supply and demand must be matched at each point and at each time on the grid. It has long been recognized that these special characteristics make the physical infrastructure of an electricity system—the grid—a natural monopoly that must be opened to all on nondiscriminatory terms before efficient and effective competition is possible. But there is an equally important natural monopoly that is not usually recognized as such: the institutional infrastructure of the dispatch process or (usually hidden) market that coordinates actions and facilitates transactions among the economic units on the grid. Trying to introduce competition in electricity supply without opening the infrastructure monopoly to all players will lead to inefficiency and frustration—as will soon be discovered in the United States as wheeling is implemented under the Energy Policy Act of 1992.

Because electricity cannot be stored, the demand for and supply of electricity must be matched at each point on the grid virtually instantaneously. Central dispatch accomplishes this matching by directly controlling generating units in real time, using technical information for each generator and offtaker to maintain system safety and reliability and, in more sophisticated applications, using cost information for each unit to minimize (approximately) the total cost of meeting demand. If this matching of demand and supply is to be accomplished by a market, buy-sell offers must be made and market-clearing prices must be determined every few minutes, for each of several (or dozens) of electricity distinct locations on the grid. The conditions of demand and supply, and hence the pattern of efficient prices and trades, change too much too quickly for trade in electricity to be efficient if it is conducted by buyers and sellers meeting occasionally to agree to bilateral contracts under which they then operate in isolation from the rest of the system—even though this is essentially the model underlying most of the discussion of "wheeling" in the United States.[2]

In concept, a decentralized electricity market could operate with individual buyers and sellers negotiating bilaterally with all others and with the grid to determine the energy prices and wheeling charges for each transaction—a process that, in a system of N entities, would involve something like N-factorial simultaneous, instantaneous bilateral negotiations, each referencing all others. However, even in concept it would be much more efficient to operate through a centralized market, analogous to centralized spot markets in other commodities. In such a market buyers and sellers make contingent buy-sell offers indicating how much they are willing to buy or sell at various prices, and then these offers are used in some centralized process to determine market-clearing prices for the system as a whole. The operator of the market can (indeed, for electricity, must) then redistribute the product and the money among the buyers and sellers whose offers were accepted, with offerers perhaps not even knowing the final outcome until after the fact.

Such centralized market processes are common in the real world, particularly for standardized commodities such as financial securities, commodities, and commodity contracts. The principal distinguishing feature of an electricity spot market is that it must work virtually instantaneously and in close coordination with the flow of the actual physical product—that is, it must be integrated with the dispatch process. If the dispatch process is doing a reasonably good job of operating the system efficiently, it need not be changed significantly; all that is required is that bidding, pricing, and settlement mechanisms be added.

The short run—say, half hourly (modern dispatch processes typically work with intervals of a half hour or less)—electricity prices that are at the center of this market and dispatch process can be more or less complex and sophisticated. The prices need not be precisely "right" to accomplish a reasonably efficient result, particularly in comparison to the far-from-efficient dispatch processes used in many existing systems. The critical thing is that all market participants can buy and sell electricity (and other services, such as reactive power, spinning reserve, and so on) at non-discriminatory prices that reasonably reflect the economic value of these services to the system at the time and place of the transaction. With such a market in place, even small, undiversified generators and middleman suppliers can compete on a level playing field with large, diversified utilities, with each player specializing in what it does best and relying on the market to provide complementary services. Without such a market, a diversified player with its own internal market will have a competitive advantage over others even if it is not particularly good at anything in particular; and only a monopoly controlling everything can capture all the benefits of trade.

2.2.3. The Role of Contracts

An electricity spot market is necessary for combining efficiency with competition but is not sufficient because, by itself, it provides no way for an individual generator to deal directly with a consumer and (the same problem stated a different way) leaves both generators and consumers exposed to the risks of the highly volatile and unpredictable spot-market prices. The solution is clearly some sort of bilateral contract allowing generators and consumers to deal with one another directly or through intermediaries—but without distorting the dispatch-based spot market that exists to minimize the total costs of meeting total demand.

The key to efficient generator-consumer contracting is an option or hedging contract based on, but not directly influencing, the outcome of the spot market. When both parties to a commodity contract have access to the same spot market, there is no need for them to have any direct physical relationship with each other or to coordinate their activities in any way. The seller can deliver its product to the market and collect the market price, while the buyer can purchase what it needs from the spot market at essentially the same price. When the spot-market price is above (below) the contract price, the buyer will pay more (less) while the seller will be paid more (less) than the contract price. All that is required to give each party what it expected from the contract is a payment between the parties that depends on the contract price and the spot market price. If the reference spot market is reasonably efficient and stable, it can be used as the basis for long-term contracts that define and allocate the risks and rewards of new investment—including even the type of contract that U.S. IPP developers have come to regard as their natural right, in which buyers take all the market risk and the developer collects a fee without putting up any real equity.

Commodity markets typically consist of one or several spot markets in which the actual commodities are traded at market-clearing spot prices, combined with markets trading contracts that are referenced to spot-market prices, allowing both buyers and sellers to operate freely in the spot market while hedging against spot-price movements. Although organized option markets trade standardized contracts for maximum liquidity, any two parties are free to enter into any contract they may agree and to settle it privately with reference to the spot-market price, with no need to tell any third party, including the spot market, about it. Such bilateral hedging contracts work even when the parties do not actually trade product with each other or even through the same physical markets; oil contracts are commonly referenced to Rotterdam or West Texas prices even though neither party may actually trade in the reference market. All that is required for an effective hedge is that the price in the reference spot market move more or less in step with the prices in the markets in which the contract buyer and seller actually operate.

By the standards of most functioning commodity markets, an electricity spot market operating on an integrated grid without transmission constraints is unusually efficient and amenable to hedging contracts because all buyers pay and all sellers are paid the same price except for a (presumably small and predictable) wholesale-retail markup covering system costs. If transmission constraints (transportation costs) divide the market into several interacting submarkets with different spot energy prices, buyers and sellers located at electrically distinct points on the grid may face very different prices, making it impossible for them to hedge prices bilaterally; but the market itself will be collecting rents on the locational price differentials and hence can risklessly offer hedges against them.[3]

The theory of spot markets and hedging contracts had not been much developed for the special case of electricity prior to the British privatization; indeed, when the British began the process, they had no such model (or any other coherent model) in mind. But they plunged in, developing and applying the theory as they went along. They made a few serious and many smaller mistakes, with predictable results that are now causing problems; but in time-honored British fashion, they muddled through and now have a system that "works."

2.3. The British Electricity Market

In the United Kingdom[4] prior to 1988, the government-owned Central Electricity Generating Board (CEGB) owned all significant generation and the national transmission grid, selling its electricity under a bulk supply tariff (BST) to regional monopoly distributors (also government owned) and to a few large direct customers. The CEGB decided how much of what to build and when, recovering through the BST all costs plus a nominal return to the government's capital. A law passed in the early 1980s allowed independent generators to sell power to the regional distributors, but the CEGB had no problem adjusting the fixed and variable components of the BST to make this option unattractive. It was a world in which many U.S. utility executives would have felt perfectly at home.

Then, in 1988, the British government issued a White Paper outlining its plans to privatize the U.K. electricity supply industry. The government knew little about the electricity business but was determined not to repeat the pattern for which it has been widely criticized in the privatization of British Telecoms and British Gas: the creation of a lightly regulated private monopoly to replace the public monopoly. So the White Paper promised to break up the industry, get the CEGB out of the business of planning generation, allow direct customer-generator contracting (at least for large customers), and maintain central economic dispatch—with the details to be worked out by the industry on a demanding timetable. In effect, the government tore the industry into small pieces, through it into the air, and told it to figure out how to put itself back together before it hit the ground. Given this start, it is amazing it has come out as well as it has.

2.3.1. Structure of The U.K. System

The U.K. electricity system now consists of the following principal elements:

1. *National Grid Company (NGC)*: NGC owns and operates the national grid, providing nondiscriminatory access and pricing; its revenue from use-of-system charges is subject to a regulatory ceiling. NGC also administers the pool's dispatch, pricing, and settlements system under contract to the pool. The twelve regional electricity companies (RECs) jointly own NGC, in the sense of having capital invested in it and expecting a return on that capital, but have no effective control over NGC.

2. *Regional electricity companies (RECs)*: The twelve regional distributors have been floated as independent RECs. The RECs must make their distribution system available to all users on a nondiscriminatory basis, allowing competing suppliers or power merchants to sell directly to customers (initially only large customers) by paying the REC's

posted and regulated distribution tariffs. Each REC's own retail supply business serves as supplier of last resort for small customers in its region but can satisfy this obligation simply by selling energy at pool prices, whatever these turn out to be.

3. *Separate, Independent generating companies*: The CEGB's generating capacity—about 50 gigawatts (GW)—was divided among three separate companies: Nuclear Electricity, which owns all nuclear plants (15 to 20 percent of the total capacity, depending on what is counted) and remains under government ownership; National Power, which owns about half of the total capacity (mostly coal-fired); and PowerGen, which owns about 30 percent of the capacity. National Power and PowerGen have been floated to the public and are now private companies. A dozen or so independent generating companies have been established and, along with National Power and PowerGen, are in various stages of building about 14 GW of new generating plant.

4. *Electricity suppliers*: In the jargon of the U.K. system, a supplier is somebody who buys electricity, either under contract or from the pool, and resells it to final consumers. RECs, as first-tier suppliers, have been granted a four-year franchise over all customers with maximum demands below 1 MW and an additional four-year franchise for customers smaller than 100 kW. But otherwise all customers are free to shop among second-tier suppliers (most of whom are affiliated with RECs or generators) for their electricity supplies.

5. *The pool*: The pool is a special legal entity controlled by generators, RECs, and independent suppliers or large customers, responsible for the electricity spot-trading arrangements that are managed by NGC under contract. The pool itself has a small staff, with an executive committee that meets to review developments, discuss problems with the trading arrangements, and suggest fixes; significant changes in the pool rules will be very difficult because virtual unanimity is required.

6. *Director General of Electricity Supply (DGES)*: The DGES's primary responsibilities are to investigate complaints, monitor and encourage competition in the industry, reset the REC and NGC price caps every five years, issue new licenses, revise licenses with the consent of the licensee, and influence capacity additions (if necessary) by adjusting the value of lost load (VOLL) parameter that influences the pricing of capacity in the pool. The DGES's principal enforcement powers come from general public and political pressure, the five-year review of REC and NGC price conditions, and the threat

to refer a licensee (including National Power and PowerGen) to the Monopolies and Mergers Commission (MMC) for possible unilateral license revision or restructuring.

7. *The nonfossil levy*: All electricity sold by a licensed supplier is subject to a levy or tax, equal to about 10 percent of the pool price, used to subsidize nonfossil sources of electricity. Although billed as a device to encourage renewable and environmentally benign electricity sources, in practice this is largely a subsidy for the nuclear program— mostly to pay the (still unknown) decommissioning costs of old nuclear power plants, now that the CEGB's plans for all-but-one future nuclear plants have been scrapped.

2.3.2. The Pool Pricing Process

The pool, with the National Grid Company (NGC) acting as the pool administrator, provides a spot market in electricity, establishing the half hourly market-clearing prices at which electricity and system services are bought and sold but taking no position in the market itself. The principal features of this process are the following.

1. *Generator offer prices*: Each day, each generating unit notifies NGC of the capacity it expects to have available each half hour tomorrow and the energy prices at which it will generate. The energy offer prices are fixed for the day ahead and include start-up costs, no-load costs, and an output-price curve with up to three linear segments. Each generator also notifies NGC of the prices at which it will provide various ancillary services such as spinning reserve and reactive power, although ancillary services are increasingly provided to NGC under longer-term contracts.

2. *The unconstrained dispatch*: NGC determines an unconstrained schedule and dispatch, indicating which units would, in the absence of any transmission constraints, be scheduled and run tomorrow to minimize system costs. This purely notional dispatch is often significantly different from the projected real dispatch, which in turn is different from the actual dispatch on the day.

3. *System marginal price (SMP)*: The SMP (in pence per kilowatt hour) for energy in each half hour tomorrow is conceptually the short-run marginal cost (SRMC) of meeting demand in that half hour in the unconstrained dispatch. In simplest terms it is the offer price of the highest-running-cost plant (notionally) expected to operate in that

half hour, although in practice the pricing process in more complicated because even the unconstrained dispatch considers start-up
costs and plant dynamics.

4. *Capacity price (LOLP/VOLL)*: Generating plant availabilities and
NGC's forecast of load are used to compute a loss-of-load probability (LOLP, defined as the probability that system voltage will be
reduced because of inadequate generation capacity) each half hour
tomorrow. LOLP is multiplied by the value of lost load (VOLL),
to determine a capacity "adder" for each half hour. VOLL was initially set at 200 pence per kWh, a level long used by the CEGB in
deciding when to add peaking capacity and roughly consistent with
values of similar variables used in utility planning elsewhere. VOLL
escalates with inflation (it was roughly equivalent to \$3.5 per kWh
in February 1993) and can be adjusted upward by the DGES if he
concludes that inadequate capacity is being added to the system.

5. *Pool purchase price (PPP)*: The half hourly SMPs and capacity price
are combined into a half hourly PPP, given (in pence per kWh) by:[5]

$$PPP = (1 - LOLP)*SMP + LOLP*VOLL = SMP + LOLP* \\ (VOLL - SMP)$$

The pool pays PPP for each kWh notionally generated in the
unconstrained dispatch, whatever the generator's bid price.

6. *Availability payments*: Each kW of generating capacity that declares
today that it will be (and then is) available during a half hour tomorrow but then is not dispatched to produce energy receives an
availability payment related to LOLP and VOLL.[6]

7. *Actual dispatch and out-of-merit Running*: After determining the
notional, unconstrained, ex ante pool prices, NGC determines an
actual dispatch reflecting transmission constraints and last-minute
changes in plant availability and system demand. Generators are
required, subject to defined penalties, to follow NGC's instructions
but are compensated for any divergences between the unconstrained
and the actual dispatch.[7]

8. *"Uplift" and pool selling price (PSP)*: PSP is calculated for each half
hour as the PPP plus an *uplift*. The uplift is the sum, for that half
hour, of all payments the pool makes for available-but-undispatched
capacity, ancillary services, out-of-merit running, and transmission
losses, all divided by the total kWh sold in that half hour;[8] certain
other costs, such as the costs of operating the pool, are included in
the uplift during daytime hours. All energy taken from the grid at
bulk supply points is sold at the PSP.

9. *Settlement of payments*: NGC uses its data on generation deliveries and supplier takes, combined with the PPP and PSP, to debit and credit the pool accounts held by each pool member. The pool takes no position in the market, simply transferring money from buyers to sellers at the end of each day and applying stringent credit controls to protect itself against nonpayment.

10. *Correction of the pool prices*: Because the ex ante dispatch and pool prices are based on information that might not be complete or accurate, and mistakes can be made in calculating the prices, the pool prices are recomputed twenty-eight days later, with final payments made on the basis of these adjusted prices. In practice, uplift is the largest source of uncertainty about the final pool prices because it cannot be known until after the fact and has been larger and less unpredictable than expected.[9]

2.3.3. The Obligation to Supply and Long-Term Contracts

Nobody in the U.K. market has any "obligation to supply," in the usual utility sense of being required to plan and acquire adequate generating capacity to meet future demand. Each REC has an obligation to connect all consumers in its region and to sell them energy at its current purchase cost (the sum of pool and contract payments/rebates) plus a small, regulated margin but has no obligation to build plants or to enter into long-term contracts on behalf of its customers, franchise or otherwise. Consumers must be willing to pay current market prices or to contract long-term if they want to get service, and new capacity will be built only when somebody—a generator wanting to sell, a consumer wanting to buy, or a speculator wanting to gamble—thinks the new capacity will provide energy and other pool-compensated services at costs that are attractive relative to projected pool prices.

Although nobody has any obligation to build or to contract for capacity long-term, anybody can enter into any kind of short-term or long-term contracts they can find somebody to sign on the other side, including the kinds of power purchase agreements that underlie most IPP financings in the United States. Contracts can be written to share risks in virtually any sensible way without interfering with the short-term efficiency of the spot market. This contractual freedom is one of the principal innovations and strengths of the U.K. system, meriting widespread study and emulation.

In the U.K. system, all real energy transactions are, in the first instance, with the pool, in the sense that the pool pays PPP for all energy delivered and charges PSP for all energy taken. But option contracts or "contracts for differences" based on the pool price can be written to protect the parties from its fluctuations. For example, supplier S can contract with a 100 MW generator G, paying G a fixed annual fee in exchange for a variable payment from G to S equal to 100 MW multiplied by the difference between the pool price and 2 pence per kWh in each half hour specified in the contract. S then buys whatever energy it needs from the pool while G sells whatever energy it profitably can into the pool, with G making the pool-price related payment to S as specified in their contract. The net effect of this arrangement is to give G a fixed income to pay its fixed costs and 2 pence per kWh to pay its energy costs (which, if G has negotiated a sensible contract, will be about 2 pence per kWh), while assuring supplier S of 100 MWh of energy each hour at a net cost of 2 pence per kWh plus the fixed contract payment (and the pool uplift). Both S and G are protected from the effects of fluctuations in the pool price but since both can still buy and sell energy at the pool price, the pool price remains the marginal cost or value of energy to each of them.

Anybody, even if not a pool member or even in the electricity business, can buy or sell financial instruments under which monetary payments depend on the pool price. Consumers can buy electricity at the pool price from any supplier and then buy from somebody else an insurance policy that pays cash compensation if the pool price goes very high; the insurer (who need not be a pool member) can then cover its risk by finding (or being) a generator willing to give up part of any upside resulting from high pool prices in exchange for a guaranteed revenue stream. Speculators can take unmatched positions in the market if they choose, although this will be at least as risky here as in other commodity markets. Neither the pool administrator nor the regulator need even know about such contracts— and a good thing, too, because they can be written and administered by the chaps down at the pub, using the pool prices published in the financial pages of daily newspapers.

The IPP projects underway in the United Kingdom are based on contracts for differences that are functionally equivalent to the power purchase agreements underlying IPP financing in the United States—that is, they transfer most risks to the buyer so that the seller can obtain highly leveraged financing. For defining and allocating the equity risks and rewards of a major project, long-term contracts are essential. However, for other purposes it is not at all clear that long-term contracts will prove essential or even useful in a U.K.-style market. Long-term contracts in other commodity markets serve primarily to ensure ongoing relationships, with price typically linked to spot market indexes. In a U.K.-style electricity market, where all buyers and sellers automatically have access to the same market and essentially the same prices, long-term contracts may have little role to play other than for financing new investments. The dominant contract in the United Kingdom may turn out to be a one-year "weather insurance" contract, which allows distributors and customers to budget for the upcoming year but leaves long-term generating asset value risk where it naturally belongs—with the owner of the asset.

2.3.4. Entry and Competition

For anyone willing to take the risks inherent in a commodity market or able to find a buyer willing to assume these risks under contract, entry into the U.K. electricity market is easy—or at least as easy as entry into such a technically and commercially complex industry should be. Anyone wanting to buy or sell electricity at the pool price can apply for a license from the DGES, join the pool, negotiate a connection agreement with NGC or a REC, and start generating or taking electricity, with no need for a contract with any other generator or supplier for energy or for back-up; such free entry can be allowed because (in concept) the pool price in each half hour reflects the true economic value of capacity and energy to the system. Everything an entrant needs to enter the market is available from entities that have a legal obligation not to discriminate and, more important, that have no interest in preventing entry (other than the RECs, which have an interest in generation and supply). This is in sharp contrast to the wheeling arrangements in the United States, where an entrant usually must negotiate complex contract terms with a direct competitor that has strong incentives to keep the entrant out of the market.

There are some limits on competition, either as transitional arrangements or as ways to discourage vertical reintegration of the industry. Nobody but the local REC is allowed to provide a supply of electricity to any premise with maximum demand below 1 MW for the first four years and below 100 kW for the next four years[10]—although anybody can get a license to build a competing distribution system! This temporary REC supply franchise was a quid pro quo for the RECs signing above-market supply contracts with the generators, who were in turn forced to sign above-market, three-year contracts with British Coal; if the protection for British Coal is extended, the REC franchise may be also. When and if the REC's de jure franchise protection falls away, even small customers will be able to get their supplies from competing suppliers. Chains of pubs, hotels, and retail stores have already sought competitive supply contracts for all their sites as a whole, and firms with easy access to customers (such us credit card companies and British Telecom) will be able to offer supply to individual households.

To discourage vertical integration, RECs are prohibited from owning generation with capacity exceeding 15 percent of their maximum demand and National Power and PowerGen jointly are prohibited from contracting directly to supply more than 15 percent of the load in any REC service territory. The DGES can alter these limits and has already increased the limit on NP/PG direct supply to 25 percent for the service areas of RECs with large industrial loads.[11]

2.3.5. Regulation

The British are intent on minimizing regulation and particularly on avoiding "U.S.-style" price regulation based on cost-plus and rate-of-return concepts. Their preferred approach to regulation is to require all entities in the industry to be subject to licenses defining their rights and obligations in broad terms and to set ceiling prices (actually, average revenue per unit of "service," somehow defined) by formula rather than trying to control profits. To take account of inflation and expected productivity improvements, the price ceiling is allowed to increase year to year by a factor referred to as "RPI-X"—the retail price index (RPI) less a target rate of real price decrease (X). The price level and the X factor are reconsidered every five years or more frequently if significant factors change unexpectedly. Variations of RPI-X regulation have been applied to the U.K. electricity industry.[12]

NGC's annual transmission revenues per unit of connected generating capacity are limited by a RPI-X ceiling; certain revenues (such as charges for user-specific connection equipment) are excluded from this ceiling. NGC collects its allowed revenues through site-specific entry and exit charges to pay for NGC equipment required to serve a specific customer; system service charges paid only by off-takers, representing the cost of a hypothetical "skeletal" grid that all off-takers would need for reliability even if they did not use the grid for bulk power transfers; and regional infrastructure charges, with demand and energy components, intended to reflect the cost of moving power from and to each region, given the current and projected locations of generation and load.

For the RECs, distribution charges are regulated by RPI-X ceilings and nondiscrimination conditions. Power purchase costs are simply passed through to retail purchasers, subject to a vague license condition to "purchase economically" and with the addition of a small supply business margin subject to a RPI-X cap. RECs have been allowed to purchase power from generators in which they have an unregulated equity interest and pass the costs through to their franchise customers, creating potential (some would say actual) abuses from self-dealing.

Generators are free from explicit regulation, beyond requirements that they not discriminate among customers, keep separate accounts, make information available to the DGES, and so on. They are required to offer ancillary services (such as spinning reserves and reactive power) to NGC at "reasonable," cost-related prices; but their pool bid prices for energy and availability are unregulated and need not be cost justified. The initial short-term (one- to five-year), government-imposed contracts with RECs limit the payoff to National Power and PowerGen from increasing pool prices; but when these contracts expire (beginning in April 1993), only fear of entry, general antimonopoly law, and the strong informal pressures that the British political system allows officials to apply will limit the ability of the National Power/PowerGen duopoly to determine pool prices and, in the longer run, contract prices.

2.3.6. The Results of Competition in the United Kingdom

The U.K. experience has demonstrated the hollowness of two bogeymen often trotted out to scare away competitive electricity markets: it is impossible to operate a competitive electricity system, so the lights will go out in the short run; and nobody will invest in new generation capacity for the long run unless there is a monopoly buyer. In fact, the U.K. system is running well (thanks to excess capacity and warm winters, mutter the critics who said the lights would go out), with significant amounts of new generating capacity under construction (too much, and only because the RECs still have some monopoly over customers, grumble those who said there would never be any investment). So criticism of the U.K. experience has shifted ground, to focus on complexity, on the behavior of costs and prices, and on some clear analomies and inefficiencies in the market.

A full evaluation of the U.K. experience is beyond the scope of this chapter and is in any case premature. There are some easy things to say: the CEGB's massive nuclear and coal construction programs have been scrapped, saving billions of dollars; the level of investment in new gas-fired baseload plant seems excessive; the generating companies formed from the CEGB (even Nuclear Electric, which is still government owned) have become much more commercially oriented and have reduced costs and improved performance significantly; the largest industrial consumers have lost their subsidized electricity supplies and British Coal is in danger of losing its protected market—and both are complaining bitterly and may yet cut a political deal to protect themselves from competition; average consumer prices have come down in real terms, although perhaps not as much as they would have under some other arrangements (such as through continuation of the CEGB but cancellation of its construction programs); industry executive salaries and profit levels are high enough to invite tabloid and Labour Party outrage; there are continuing complaints about the complexity and high overhead costs of the system; and there are more than enough design and implementation errors in the system that those who said it would never work can save some face by saying "I told you so; just you wait and see."

It will take some years before a full evaluation of the U.K. system is possible, and even then it will be inconclusive. What can be said with certainty is that some serious mistakes were made in the United Kingdom's competitive privatization and are having predictable consequences. But the most serious errors have little to do with competition as a concept or even with the specific market arrangements, imperfect as they are. Two features of the system are particularly noteworthy, neither having to do with market mechanics as such: There are two generating companies owning 50 percent and 30 percent, respectively, of the generating capacity; and the RECs are allowed to buy power from their own unregulated generating subsidiaries and pass the costs through to captive customers. Given these structural features, it is no surprise there are complaints of inadequate competition and of self-dealing by the RECs, whatever the merits or lack thereof of the market mechanism.

As costly as the problems of inadequate structural competition in generation and inappropriate REC regulation may be in the short run, they should be temporary and self-correcting because the market is doing more or less what a market is supposed to do. High duopoly pricing is encouraging entry that will eventually erode the duopoly, and the problem of REC self-dealing will fade away as the 15 percent limit is reached and the franchise is phased out. But there are some serious distortions in the market itself that should be corrected in the United Kingdom and that can be avoided in implementing competitive concepts elsewhere. There is as much to be learned from the failures of the U.K. experience as from its successes.

2.4. Improving the U.K. Model

The U.K. electricity market suffers from three principal problems, in addition to the basic structural and regulatory flaws mentioned above: (1) its energy/capacity pricing rules are largely ad hoc and hence too complex and easily "gamed"; (2) pool prices are fixed in advance and on the basis of a fictional unconstrained dispatch and administrative LOLP calculation rather than with a true market-clearing process; and (3) transmission is not properly integrated with the pool, creating serious pricing and investment distortions when transmission constraints are important. These interrelated problems and the outlines of the solutions are discussed below.

2.4.1. Simplifying the Pricing Process

An electricity spot market should reproduce the results of a good economic scheduling and dispatch process, such as (improved versions of) those used in sophisticated power pools. If a scheduling/dispatch process is producing reasonably efficient day-to-day operations and a proposed spot-market process would significantly change these results, the proposed spot-market process should be reconsidered—along with the dispatch process itself. The key to establishing a workably competitive and reasonably efficient electricity spot market is to integrate the market and economic dispatch into a single process.

In the simplest model of an economic dispatch process, all available generating units are ranked by increasing energy cost to create a system marginal cost or merit order curve, and plants are dispatched in order of increasing energy cost until demand is met. It is easy to convert such a process into a market, simply by allowing each generator to bid an energy price and a quantity, accepting bids in order of increasing price, setting the system marginal price (SMP) equal to the last bid accepted, and paying SMP for all energy taken. Unfortunately, in reality it is much more complex.

Dispatching a complex electricity system and setting prices are made difficult by several technical realities. For one thing, generating plants are big machines with inherent inflexibilities, such as limits on the frequency or rapidity at which they can be brought to full power or shut down. For these reasons, a high-running-cost plant operating at partial capacity at night may not be the marginal generator at night but may be waiting to be run up to meet the rapid growth in demand the next morning; in this case, the marginal cost of meeting load at night is determined by some other plant, and the costs of running the high-cost plant at night should be allocated somehow to the morning hours. Similarly, planned maintenance of generating plants and transmission lines must be scheduled months or even years in advance, and units must be commited to various stages of readiness days or weeks in advance, to ensure that adequate generating plant will be available to meet load on the day. These commitment and dispatch decisions are inherently "lumpy" and interdependent and hence do not lend themselves to decentralized market processes; somebody must consider all the possible combinations of plants that can be scheduled and operated, choose the best (or at least a "good") combination for meeting demand, and determine a set of prices based on this dispatch.

In concept the prices based on the dispatch should equal SRMC, the increase in system cost necessary to meet a one unit increase in demand for energy (or other service) in a specific half hour. However, the interdependencies and lumpiness of the scheduling and dispatch problem make it difficult, even in principle, to define SRMC unambiguously, much less to measure it, so some pragmatic rules must be used to determine prices. The prices and dispatch should meet at least two basic criteria: (1) the outcome prices and quantities should be consistent with the buy-sell offers of market participants, in the sense that the market should clear and each bidder should be satisfied with what it ended up buying and selling at the final prices, even if it had no real-time control over these amounts; and (2) the prices should provide "the right" incentives for efficient short-term operations and long-term investment.

Determining a dispatch and prices that meet these criteria even approximately is not easy but can be done in a logical way using modern techniques. Unfortunately, the pricing rules in the United Kingdom were negotiated one at a time, in a highly charged adversarial process, under severe time pressure, starting with a relatively unsophisticated and inflexible dispatch process, by people who were discovering the principles and inventing the procedures as they went along. It is remarkable that the result is no worse than it is; but it is not a pretty sight. The complex and ad hoc pricing rules create many anomalies and opportunities for inefficient gaming and are a major reason for the complaints about the complexity of the U.K. system. Much simpler and more efficient dispatch and pricing processes can and should be developed.

The U.K. pricing system is regarded as complex largely because nobody knows what it is trying to do or what it actually does and because it makes payments for all sorts of strange things and provides (or at least appears to provide) opportunities for clever people to make money by gaming the system rather than by offering valuable services at attractive prices.[13] Moving to the ex post pricing arrangements discussed below would reduce many of these problems by reducing payments for reserve, forecast errors, and last-minute changes. And a more automated and scientific pricing process can be developed, allowing the dispatchers to exercise the human judgments that are always required but computing a set of prices based on more consistent and readily understandable principles.

For example, if the pricing problem were defined in terms of a logical cost-minimization problem, taking as constraints the factors identified by the dispatcher on the day, prices could be determined in a mathematically sophisticated "black box" computer program. It would then be relatively easy to explain what the pricing system is doing, and there should be a clear relationship between demand, supply, bids, and the resulting pool prices. There would be less scope for gaming by looking for gaps or discontinuities in the pricing equations, and the prices would be less easily influenced by individual bidding behavior. Such a pricing process would probably appear to be much less complex, even though the mechanics of the solution algorithm are understood by nobody but the programmer. Of course, a generator controlling half the capacity on the system could still withhold capacity or submit high bids to increase prices; but these prices would be clearly seen to be the result of market power, not the "complexities" of the pricing process.

2.4.2. Ex Post Pool Pricing Based on the Actual Outcome

The U.K. system bases pool prices on a notional, ex ante, unconstrained dispatch that may bear little relationship to actual system operations. This is not only illogical but has created some predictable problems. When generators know that their bids will affect the pool price but not their actual dispatch and payments, or vice versa, they have incentives to bid strategically. For example, they can withhold capacity in their day-ahead bid to drive up LOLP and then redeclare the capacity available the next day to collect the high availability payments. Or a plant downstream from a transmission constraint can bid high, knowing that it will not be included in the unconstrained dispatch but that in the actual dispatch it will be run and paid its high bid. Even if there is no conscious gaming, the incorrect price signals can produce inefficiencies, as when a consumer takes costly load-reducing actions in response to a high LOLP announced a day ahead, but then there is no need for such action on the day because a generator comes back on line unexpectedly.

As such problems have been identified in the United Kingdom, they have been investigated by the DGES and various ad hoc fixes have been applied. But as long as pool pricing is based on fiction rather than reality, there will be no end to such problems. The only real solution is to develop a pool pricing process that prices energy and capacity on the basis of what really happens—which means, in practice, that final transaction prices may not be known until after the fact.

Although the suggestion to determine final transaction prices after the event often sounds strange and illogical on first hearing, in fact it is quite common and sensible. Most decisions in the world, from building an oil refinery to going to the store for a loaf of bread, are based on projected prices, not certain prices. In commodity and financial transactions it is routine for a buyer or seller to tell a broker to buy or sell something on the best possible terms, perhaps subject to a reservation price, but with the buyer or seller not knowing how much "excess profit" was made relative to the reservation price until after the deal is done. In electricity, U.S. power pools compute transaction prices and settle payments after it is known how the system was actually operated. In the U.K. pool now, most of a generator's costs are committed days or weeks or months before the day-ahead prices are announced; the fact that a generator is not certain about Wednesday's prices on Tuesday would be a minor increment on the uncertainty a generator now faces about pool prices next Thursday or next week when deciding whether to commit its plant to a two-week run or a two-month overhaul. Similarly on the demand side, any consumer for whom tomorrow's price at 6:00 P.M., is critical can either contract today for a known price tomorrow or can give instructions to throw the switch at 5:00 P.M. tomorrow if it looks like the price an hour ahead is likely to exceed some critical value. As long as market paritcipants can make contingent offers or can contract against the pool price, there is no serious cost or inconvenience created by being uncertain about tomorrow's final pool price.

An ex post pool pricing process with a round of ex ante contracting might work something like the following:

1. *Generator offer prices for energy and ancillary services*: Each generating unit submits location-specific offer prices and availabilities on a day-ahead (or perhaps week-ahead) basis, for both energy and ancillary services such as reactive power. These offers are regarded as firm, in the sense that once they are accepted the generator will be expected to deliver the agreed service at the agreed price or pay the consequences.

2. *Demand-side offers to buy energy*: Suppliers and large consumers submit location-specific bids to buy energy, indicating how much energy each is committing to take during each half hour over the dispatch period, with increments or decrements from the projected amounts at various prices—such as, 5 percent less in any hour in

which the pool price exceeds 5 cents per kWh or 10 percent more if the pool price falls below 2 cents per kWh.

3. *Ex ante constrained dispatch and nodal prices*: Using the day-ahead (or week ahead) offers, the pool schedules generation and ancillary services to meet projected demand at least cost, taking into account expected transmission constraints, and then calculates half hourly SMPs for each of the electrically distinct regions or nodes in the system. In this process load management bids are treated much like generator bids, in the sense that SMP is increased as higher-priced load management increments are scheduled. The outputs of this initial ex ante calculation are the expected values of pool prices, the output of each generator and the offtakes of each supplier, as these will be calculated ex post.

4. *Contracts based on the ex ante prices and quantities*: The ex ante expected prices and quantities determined by the pool can be regarded as indicative measures, with generators and offtakers free to contract bilaterally with one another for the following day. Alternatively, they can be regarded as firm contracts between generators and the pool, and between the pool and offtakers, with the pool in a back-to-back or fully hedged position. For example, if the ex ante dispatch says that generator G will produce 1,000 MWh at a price of 30 pounds per MWh, generator G will have to produce 1,000 MWh if called on to do so in the final dispatch or will have to buy any shortfall at the ex post prices; and the pool will have to pay G for lost profits if G is not dispatched. The pool should have a matching commitment from supplier S to buy 1,000 MWh at a price of 30 pounds per MWh (plus uplift) and to buy or sell any incremental take at the *ex post* price.

5. *Final dispatch, ex post prices, and settlement*: The pool updates estimates of plant availability and demand, adjusts the planned dispatch and estimated pool prices accordingly, and announces the results as often as practical. At the end of the day, after load has been met (or not) in each region, the ex post, nodal SMPs are computed, and settlement is made at these prices, either on all energy actually delivered and taken or, if the pool was a party to any ex ante contracts, on the differences between ex post and ex ante quantities.

6. *Uplift*: The uplift should be relatively small because it will not include the costs of transmission losses, out-of-merit running, or differences between ex ante and ex post dispatches. However, there may be some ancillary services and pool overhead costs to be recovered through an uplift of the selling price over the buying price.

7. *Pricing generation capacity*: With ex post pricing there would be no need for a LOLP calculation; if capacity were not adequate to meet demand without voltage reductions the ex post SMP would be set at some high value analogous to the VOLL in the U.K. system. However, with demand bidding and a high VOLL there should seldom be involuntary demand reductions because voluntary load management would be called at increasingly higher SMPs until the market cleared. The ex ante price determination might include a LOLP-type calculation to estimate the expected value of SMP; forward contracts made on the basis of this expected value would play the role of the LOLP/VOLL calculation in the current U.K. pool.

8. *Transmission contracts*: As discussed in the next section, with nodal pricing the pool will collect rents on the locational price differentials. These rents should be rebated by the pool to the holders of defined transmission rights, which can be done automatically through the settlement process.

2.4.3. Integrating Transmission into the Pool

For a successful competitive electricity industry the most critical structural feature is separation of the monopoly functions—transmission[14] and dispatch or pool—from the competitive activites of generation and supply. The U.K. privatization got this split more or less right, but made a serious mistake on another split: the division of responsibilities between the transmission entity and the pool. The error was to give NGC responsibility for the fixed costs of the grid (recovered through use-of-system costs limited by a RPI-X revenue ceiling) while leaving the pool responsible for transmission losses and out-of-merit running costs (recovered through the uplift).

As a result of the inappropriate division of responsibility between NGC and the pool, the current U.K. system has no good way to evaluate or make transmission investments to lower losses and out-of-merit running costs. If NGC invests to lower losses and out-of-merit costs, pool members as a whole gain through lower uplift payments,[15] but NGC has no way to be repaid because its revenue ceiling is based on the amount of generating capacity connected to the grid, not on the value of the grid to users. This has created a situation in which NGC has strong incentives not to invest to expand grid capacity even when it is cost effective to do so. NGC is even being accused of using its use-of-system charges to induce generators to locate or close plants in ways that result in higher system operating costs but allow NGC to postpone grid investments that are cost effective from a system perspective.

Other than intrusive and ultimately unproductive regulation to try to force the grid entity to do what is not in its interest to do, the only solution to this problem is to make a single entity responsible for both grid fixed costs and system operating costs so that proper tradeoffs can be made. To illustrate how this might be done, it is useful to think in terms of three different functional organizations: Wireco, Poolco, and Transco. Wireco, a profit-seeking company, owns and maintains the physical transmission assets and leases them to Transco for an agreed lease payment, subject to defined penalties for failing to maintain agreed technical standards. Poolco is a nonexclusive, nonprofit club of generators and suppliers that operates the pricing and settlement system. Transco is a profit-seeking company that is responsible for the actual dispatch and its associated transmission losses and constraints, for making the lease payments to Wireco, and for determining when and where investments in the grid should be made; Transco's revenue is fixed, subject to a RPI-X cap, and comes from Poolco, who collects it through charges on grid users.[16]

In the day-to-day operations of the system, Poolco determines trading prices on the basis of a notional ex post dispatch of the offers of the generators and suppliers active on the day, assuming a defined baseline configuration of the grid; this baseline configuration (which could be different at different times, to reflect planned grid maintenance, or could be an unconstrained configuration at all times, as in the United Kingdom at present) is the system that Transco is responsible for maintaining. Transco then determines the actual dispatch of these same plants and bids to meet the same load, but taking into account the actual configuration of the grid on the day. Transco pays for generation and load-management costs in excess of the costs associated with Poolco's notional dispatch; these are the costs of losses and out-of-merit running due to degradation of the grid from the baseline conditions assumed in Poolco's notional pricing dispatch.[17]

Transco keeps as profit any difference between its revenue (paid by the pool subject to a RPI-X cap) and its costs—the sum of lease payments to Wireco and its payments for losses and out-of-merit running. Transco can pay Wireco (or anybody else) to invest in the grid but does not thereby get any increase in allowed revenue from grid users; such an investment is profitable for Transco only if it costs less than the system losses and out-of-merit running costs it saves. This gives Transco the right incentives to dispatch the system efficiently, to obtain good performance from Wireco, and to pay for grid improvements that are cost effective from a system perspective.

The proposed functional split of market, dispatch, and transmission functions can be used to implement nodal spot pricing and transmission contracts.[18] Poolco's notional pricing dispatch will determine the nodal spot prices implied by the baseline transmission constraints. Poolco will collect significant and volatile rents on the price differentials but will rebate these to specific grid users (or other parties) that have contractual rights to the transmission capacity between specified points, consistent with the baseline configuration of the grid. The allocation of the transmission rights should be made in conjunction with the determination of use-of-system charges; those who pay the most for the system should get the most value from it, in the form either of favorable local pool prices due to grid investments or of transmission rights compensating them for unfavorable local pool prices.

If actual transmission constraints on the day are different from the baseline constraints, Transco will instruct some generators or offtakers to do something different from the pricing dispatch and will compensate them for the resulting excess costs or lost profits. Transco has incentives to find a combination of out-of-merit running, load management, and grid operating or investment costs that responds to transmission constraints in a least-cost way. Although this would be a major improvement over the current situation in the United Kingdom, where NGC has little incentive to dispatch efficiently and has a positive disincentive to invest in the grid, bilateral arrangements are not as effective for stimulating cost-effective responses to grid constraints as proper nodal spot pricing would be; giving all generators and offtakers the right incentives to respond to grid constraints would be better than relying on Transco to find and negotiate the least-cost responses. In practice, the actual dispatch may always be somewhat different from the pricing dispatch, so there may be need for the kind of split between Transco and Poolco suggested here; but the objective should be to make the pricing dispatch match reality as closely as possible.[19]

Negotiating and enforcing the lease between Transco and Wireco will be difficult, so there is a strong argument for putting them both into a single business (as in NGC). The problem is that the resulting company will be very powerful and probably unresponsive to the generators and suppliers who use the system. Transco's revenue ceiling cannot realistically be determined for all time, even with indexing much more sophisticated than a simple RPI-X, so it will have to be adjusted periodically. This will mean that, in the long run, an asset owning Transco will be a cost-plus operation, tending—like all regulated monopolies—to goldplate its system and to be relaxed about its costs.

A solution to this problem is to make Poolco responsible for actual dispatch and management of the grid because Poolco, as a nonprofit organization with no grid assets to protect and enhance, would presumably seek to minimize its costs on behalf of its members. But an industry club (some would say cartel) never has impressive decision making abilities or cost reducing incentives. Design of the institutional arrangements for managing the natural monopoly functions that are at the core of an integrated electricity system is a difficult problem with no perfect solution.

2.5. Conclusions

With special mechanisms to deal with the special technical characteristics of electricity, the theory of competitive spot and hedging contract markets can be applied to an electricity system. The system that has been operating for several years in England and Wales applies this theory imperfectly and is having predictable problems as a result. Nonetheless, the U.K. system is working, at least in the sense that the lights are on and investment is taking place, much to the dismay of those who said it could never happen.

It will be more difficult to create effective and efficient competition in the United States electricity industry, where institutional arrangements are much different from those in the United Kingdom prior to the competitive privatization. But the drive toward competition has begun and appears to be unstoppable in the United States. U.S. utility executives and regulators would be well advised to study both the theory and the experience developed elsewhere before rushing ahead with plans for wheeling and open access that are often based on a poor understanding of either the theory or the practical problems involved in a competitive electricity market. If the wheeling model proceeds without adequate development of the spot market and contract arrangements necessary for effective and efficient competition in any market, the result will be disruption, large inefficiencies and the frustration of the drive toward real competition.

Notes

1. In the United States an important distinction is made between a utility's wholesale customers—other utilities that buy for resale—and retail customers or final consumers: the Energy Policy Act encourages transmission access or "wheeling" that creates competition for wholesale customers but forbids the Federal Energy Regulatory Commission from ordering wheeling for retail customers. This distinction is unimportant for the discussion here because the same basic institutional changes are required if competition for either type of customer is to be effective and efficient. The real difference is quantitative: the U.S. utilities whose

excess and too costly generation plant puts them at grave risk from competition sell most of their power directly to final consumers, not to resellers, so limiting competition to resellers reduces the immediate threat to them. Once wholesale wheeling becomes a reality, retail wheeling will not be far behind.

2. The U.S. style wheeling model thinks of a generator G and a customer C agreeing to a defined flow of electricity at an agreed price, paying the wheeling utility a fee to move the electricity from G to C, and then operating under the contract with little or no regard for the larger system. Even if such independent operation were technically possible without endangering system safety and reliability, it would be grossly inefficient because it would deny G and C the very trading efficiencies for which integrated electricity systems were developed.

3. Transmission pricing and hedging contracts are discussed further in Section 2.4.3 of this chapter and at length in Chapter 8.

4. Actually, in England and Wales. Politically, the United Kingdom consists of England and Wales plus Scotland and Northern Ireland. The Scottish system has been privatized and is connected to the England and Wales system but operates somewhat differently. The Northern Ireland system has been broken up and privatized, but with Northern Ireland Electric remaining as a monopoly buyer and reseller. What is referred to here as the "U.K. system" is actually the England and Wales system.

5. PPP is a probability-weighted average of SMP and VOLL and hence can be thought of as an expected value. Roughly speaking, system SRMC tomorrow will be SMP if load is not lost, an event with probability (1-LOLP), and will be VOLL if load is lost, an event with probability LOLP. Thus, PPP is the day-ahead expected value of tomorrow's SRMC. It is perfectly logical to use an expected value of SRMC as the basis for trading futures in tomorrow's electricity; but actual trades tomorrow should be based on actual SRMC's tomorrow.

6. The logic here is the same as that justifying an annual capacity payment for a peaker that turns out not to be needed during the year; the time period is just reduced to a half hour, giving more precise and less predictable price signals. The specific formula used for availability payments in the United Kingdom makes a logically incorrect distinction between plant that runs and plant that does not run; but the effects of this are not significant in practice.

7. For example, a generator run *out of merit* because of a transmission constraint is paid its bid price, while one not-run because of a transmission constraint is paid its *lost profit* defined as the difference between the pool price and the generator's bid price. The gaming opportunities created by this arrangement have led to predictable problems. Ways to reduce these problems are discussed presently.

8. This averaging mutes some potentially important marginal cost signals. Even if the average cost of, say, transmission losses and out-of-merit running due to transmission constraints is small, it is the potentially much larger marginal cost that should be signaled through prices. Even if uplift is "only" 10 percent of the pool price, reducing its costs by, say, half would save a lot of money. The high cost of the uplift is due largely to the artificial dispatch used for pricing purposes.

9. The twenty-eight-day-later correction does not adjust the ex ante prices to reflect what actually happened in the event, but only to reflect what should have been projected a day ahead given what should have been known a day ahead. Thus, generators and consumers base their decisions, not on their best estimate of the actual outcome on the system, but on their best estimate of the outcome of a calculation that will not be completed until twenty-eight days after the fact, and then will be based on counterfactual assumptions.

10. "Maximum demand" and "premise" are both subject to wide interpretation; the DGES has already ruled that commonly owned sites served by different meters can be a single premise if they are "close together."

11. In any case, a generator can easily get around this limit by contracting with a third party that contracts with final consumers. Such transactions are neither prohibited by license nor possible to prevent or even track in this market, unless the generators are prohibited from contracting altogether.

12. The distinction between U.K.-style RPI-X regulation and U.S.-style rate-of-return regulation may be less than meets the eye. In either case, maximum prices are set on the basis of expected costs and acceptable profits over some future period, with the utility free to make more or less by controlling its costs during the period of "regulatory lag" between rate cases (in the United States) or price reviews (in the United Kingdom). When U.S. regulators index maximum prices so that the period between rate cases can be extended they are using a form of RPI-X regulation.

13. Actually, the problems of this sort are not as important as folklore would have it. The pool is doing reasonably well what it was designed to do. It has become fashionable to criticize the pool rules as complex and to mutter about LOLP/VOLL and other arcane matters, particularly among those who are really unhappy about something else, such as the generator duopoly or loss of a previous subsidy.

14. Low-voltage distribution is also a natural monopoly that may be, and outside the United States usually is, separated from transmission. There are good managerial and competitive reasons for such a separation; but the distribution entities will be regulated monopolies themselves, so whether they are divisions of the transmission entity or separate corporate entities is not critical.

15. Because the transmission-related component of uplift is paid to generators, they may have some interest in a higher uplift. However, if uplift payments were based on generator costs, as they are supposed to be, generators would just break even on uplift payments and consumers would lose. In practice, generators do gain something from uplift but not nearly as much as consumers lose.

16. Grid users and pool members are essentially the same. Pool membership may be compulsory, as in the United Kingdom; but if the pool and Transco are operating and pricing properly there will be no need to make pool membership compulsory because anybody connected to the grid will want to participate in the pool.

17. If actual generation costs are less than those determined in the pool's notional dispatch, it must be because the grid is in better condition than assumed and Transco will keep the difference.

18. See Chapter 8 for a more complete exposition of these concepts.

19. If Transco makes an investment that improves the transfer capability of the grid, it can pay for it in one of two ways: Leave the baseline grid configuration unchanged and make money from the reduction of actual generation costs below the notional costs in Poolco's pricing dispatch; or negotiate with Poolco to change the baseline configuration in exchange for an increase in Transco's RPI-X revenue cap. If Poolco agrees with Transco that the investment is worth making, it should agree to pay for it (recovering its costs from grid users through increased use-of-system charges and sale of the new transmission rights) and modify the baseline configuration accordingly because this will lead to better pool price signals.

3 THE BRITISH ELECTRICITY EXPERIMENT*

John Vickers and George Yarrow

3.1. Introduction

The privatization of the electricity supply industry (ESI) in Britain , which began 1990, is in itself an unremarkable event: much of the ESI is privately owned in several European countries and in the United States, and the U.K. privatization program has embraced utilities since 1984, when British Telecom was privatized. The thing that *is* remarkable about recent policy toward the industry is the radical nature of the regulatory reform accompanying privatization. A vertically integrated, administered generation and

* This chapter is an abridged version of Vickers and Yarrow (1991) plus a postscript on some subsequent developments. The research was part of the project on the regulation of Firms with Market Power under the UK Economic and Social Research Council initiative on the Functioning of Markets. Financial support from the ESRC and the Office of Fair Trading is gratefully acknowledged. The views expressed in the paper are entirely our own, and we are responsible for any errors. We are very grateful to Billy Jack and Kaiser Kabir for their research assistance and to Nils-Henrik von der Fehr, David Begg, Paul Seabright, Jean-Paul Rochet, and seminar audiences at Harvard University and the Massachusetts Institute of Technology for their helpful comments and suggestions.

transmission structure was replaced overnight (on March 31, 1990) by one that is deintegrated and market based: generation and transmission have been separated, and there has been an attempt to establish competition in generation. A number of countries have been moving gradually in the direction of encouraging competition, but nowhere else have such dramatic changes been implemented so quickly.

Given the importance of the industry, these reforms are of great interest for their own sake and, we suggest, for far more than that. The experiment promises to be highly informative for policy makers elsewhere who are currently assessing options for regulatory reform not only in the energy sector but in a wide range of industries. Indeed, being subject to a variety of market failures ranging from natural monopoly to environmental externalities, and with its "vertical" and "network" characteristics, the ESI provides a rich case study in public policy toward industrial organization. The radicalism of British reform brings into sharp focus a combination of pervasive issues about ownership, competition, entry conditions, vertical integration, contracts, investment incentives, price regulation, quality assurance, and environmental protection.

With these generic issues in mind, we aim in this chapter to give an interim review of the competitive and regulatory regime that has been established for the British ESI. (The regime is described in section 2.3 of the previous chapter in this volume by Larry Ruff.) We identify strengths and weaknesses of the new regime, and the key factors—including future regulatory policy—on which its success is most likely to depend. In addition we link our analysis to questions facing European energy policy more generally and draw conclusions relevant to policy issues in other industries that share some economic characteristics with the ESI.

The five sections of this chapter are devoted to a series of generic issues that arise in various industries: vertical supply arrangements (Section 3.2), oligopolistic competition and collusion (Section 3.3), entry and access terms (Section 3.4), price regulation (Section 3.5), and environmental regulation (Section 3.6). Potential lessons for energy, competition, and regulatory policies are summarized (Section 3.7). Finally, there is a postscript (Section 3.8) on some recent developments in the industry.

3.2. Vertical Supply Arrangements

The most radical features of the reform of the ESI in Britain are the vertical separation of generation from transmission, the creation of a market for wholesale power supplies, and the (partial) vertical separation of distribution and supply. These structural reforms, which are aimed chiefly at promoting competition where competition is feasible, raise questions concerning the tradeoffs between markets and hierarchies (Williamson, 1975) in a particularly clear way. Consideration of vertical supply arrangements is therefore an obvious starting point for analysis of the ESI, and, as we show, it leads on naturally to questions of competition (Sections 3.3 and 3.4) and, where effective competition is not feasible, to questions of price regulation (Section 3.5).

Our discussion of vertical issues is in two main parts. First, we abstract from network characteristics, by treating transmission as a homogeneous input to electricity supply, in order to focus on generic problems of investment, anticompetitive behavior, and contracting. Second, we examine some network economics of transmission, which have an important bearing on the question of vertical organization because of the difficulty of efficiently pricing transmission in a decentralized regime. These features are not unique to electricity—they arise also in other network industries—but they occur in their sharpest form in the ESI because of the nature of the product. Finally, we link these two sets of issues and draw some implications for policy toward competition in generation.

3.2.1. Vertical Integration

3.2.1.1. Investment Problems. In respect of the relationship between a supplier and a buyer, consider the question of why they might wish to form a long-term relationship, or integrate completely, rather than deal spot. One important reason has to do with the hold-up problem that can arise when there are sunk costs (see Tirole, 1988). If the supplier has to invest in specific assets which are much less valuable in other relationships, it faces the risk of being exploited by the buyer ex post. Confronted with that risk, the supplier might not invest in the first place.

Of course, in this example it is not a very sensible institutional arrangement to give the buyer the power to determine price ex post. But the problem can arise also in other settings. For example, if price is determined by ex post bargaining, then the supplier will have an inefficiently low incentive to invest because it will gain only a proportion of the extra value created by additional investment but will pay all of the extra cost. The hold-up problem can be solved by integration between buyer and supplier or by a contract that settles terms in advance (see Joskow, 1987).

The new structure of the ESI is, of course, much more complex than in the simple story above. There are sunk costs in both generation and transmission, there is regulation of transmission and distribution charges, and there is some competition on both sides of the wholesale market. Insights from the story are relevant nevertheless. For example, generators might be deterred from investing efficiently in plant with sunk costs for fear of exploitation by the monopsonistic grid downstream. Regulation seeks to alleviate this problem by controlling transmission charges, but there could still exist a monopsony problem further downstream—with the RECs—if they acted collusively. The more competition there is among RECs (and other suppliers) for wholesale power, the smaller this problem becomes because none can then hold the generators hostage. Put another way, competition reduces the specificity of buyer and seller relationships and serves as an alternative antidote to any underinvestment problems.

A parallel issue is raised by privatization itself. Under state ownership, investment and pricing functions are integrated. After privatization, investment decisions are made by private agents, but the state retains certain regulatory powers, especially concerning prices. The danger that a regulator or government in the future might tighten prices once investments had been sunk could be a deterrent to efficient investment: private-sector discount rates could include a premium to reflect this risk. The question is how well the government can commit itself, and its successors, not to behave in such a way (see Vickers and Yarrow, 1988; Gilbert and Newbery, 1988). Note that in the ESI the mix as well as the level of investment could be affected by these problems, which are greatest for plant that is capital intensive and has long lead times (particularly nuclear).

3.2.1.2. Risk Sharing.   Another reason for longer-term relations, including integration, between upstream and downstream parts of the ESI is risk sharing. This is not just a question of risk aversion. A generator reliant on spot market sales runs the risk of bankruptcy (and its associated costs) if, for example, demand a decade hence turns out to be lower than expected. Supply companies, on the other hand, are well placed to bear these risks because they can often pass them on to consumers. Or if they have contracts to supply consumers at fixed prices, they can avoid spot market risk themselves by contracting long-term with generators or more simply integrating with them. In the new regime in Britain, the RECs can integrate backward into generation but only to the extent of 15 percent of their needs.

3.2.1.3. Anticompetitive Conduct.   The main problem with vertical integration is that it can facilitate anticompetitive conduct. Long-term contracts can also be barriers to entry (see Aghion and Bolton, 1987), but here we focus on full integration. With integration between generation and transmission, the dominant firm might try to make access to the naturally monopolistic grid difficult or impossible for rival generating companies, thereby extending its market power from transmission to generation. A similar issue arises between distribution and supply. Vertical separation removes the incentive for such behavior, which is, of course, the main idea behind the structural reforms in the British ESI.

3.2.2. Network Issues

The points so far discussed would arise even if transmission were a homogeneous product. Further complexities, involving spatial pricing, result from the network characteristics of electricity transmission. With vertical integration of generation and transmission, the system can be optimized internally, but with separation it is necessary to have actual (rather than just shadow) transmission prices, and these have real effects. Principles of optimal spot pricing for electricity are derived by Bohn, Caramanis, and Schweppe (1984), from whom we draw in the following (see also Wilkinson, 1989).

The four characteristics of electricity supply that affect optimal pricing in their analysis are (1) transmission losses, (2) capacity constraints in the network, (3) the need for continuous electrical equilibrium at each node, and (4) the fact that electrical flows cannot be directed along particular lines because they are allocated by nature according to Kirchoff's laws. As a consequence there are major externalities across space: the optimal price at node i can be much affected by events at far away node j.

The optimal spot price at node i has three components. The first is the cost of the marginal generating unit in the system (plus any premium needed to curtail demand to the available capacity). This term, which is the same throughout the system, is then grossed up (or down) by the second component, which reflects incremental transmission losses caused by demand at node i. The reason that this second factor might be less than one is that demand or supply at node i might reduce losses in the system as a whole. Third, terms reflecting transmission constraints are added: there are shadow prices for capacity on lines that are fully loaded, and the price difference between nodes is wider than it would be if spare transmission capacity existed.

Because power flows are allocated throughout the system naturally, the (optimal spot) transmission charge at any node depends on events throughout the network and vary over time and randomly as supplies and demands alter. It is entirely possible for the optimal transmission charges at some nodes and times to be negative—that is, for a generator to be subsidized for inputting power or for a consumer to be subsidized for drawing power from the system.

These results have important consequences for competition between generators. Bohn, Caramanis, and Schweppe (1984, p. 371) examine how much market power a generator has in terms of its ability to affect the price that it faces by varying its supply. They conclude that "At some times, a generator may have no effective competitors and thus considerable power to affect prices. At other times, the same generator may find itself competing with generators hundreds of miles away. The stronger the transmission system, the more effective competition will be." This last point bears emphasis. Transmission charges are a kind of transport cost. High transport costs reduce the geographic scope of rivalry. With a stronger transmission system, capacity constraints are less likely to bite, transport costs are lowered, and competition is made more effective. Policy implications of this "procompetitive externality" of grid investment are discussed below.

3.2.3. Vertical Arrangements: Conclusions

The discussion of spot pricing can now be related to earlier observations about investment and anticompetitive behavior. First, if transmission charges are regulated according to the principles of optimal spot pricing, and if the grid operator is a profit-maximizer, then there could be serious underinvestment problems. Optimal prices fall as the grid is strengthened, and so the grid company would hold back on investment. This factor and the procompetitive externality of grid strengthening for competition in generation suggest that regulation of grid investment is important. Given the difficulties of regulating the investment of a private firm, this might be a reason for having the grid under public ownership.

Second, if vertical integration between transmission and generation is allowed, the grid company will have incentives for anticompetitive behavior in its grid investment policy (even if grid pricing is regulated according to marginal cost principles). The firm could shield its own plant from competition and raise rivals' costs. A similar point applies to international competition. Grid investment (or rather underinvestment) can be used as a means of protecting domestic suppliers. For example, a major bottleneck between France and Germany could limit the competitive threat of Electricité de France (EdF) to the German electricity (and coal) industries. This raises the question of whether a supranational Eurogrid is needed to have a truly competitive internal market in Europe.

3.3. Oligopolistic Competition

Oligopoly or oligopsony characterize several stages of production in the electricity industry. Given, however, that the major structural reforms were motivated largely by a desire to increase competitive pressures in electricity generation, we focus here on competition in the wholesale power market. Very roughly, horizontal restructuring can be viewed as a means of creating immediate competition among incumbents, and vertical separation can be viewed as a means of reducing entry barriers. The discussion is therefore divided into analysis of competition among incumbent generators (this section) and of competition from (actual and potential) entrants (Section 3.4).

3.3.1. Competition and Collusion

A full analysis of oligopoly issues must take account of a whole range of features of competition in the wholesale market. Nevertheless, it is useful as a first step to focus on a situation in which generators sell into a completely unregulated spot market with price-taking consumers.

Indeed, suppose for a moment that the industry is also perfectly competitive. Being unable to influence market prices, profit-maximising generators will offer to supply power at their marginal cost of their generating units. The result of aggregating these bids will be an industry supply curve. Price is determined by the intersection of that curve with the demand curve. Figure 3.1 shows how price varies with demand in this situation. In the top right quadrant is the annual load curve, in which half hours are ranked in decreasing order of demand. Thus demand exceeded Q megawatts during H half hours in the year. The industry supply curve, derived from marginal costs as described, is shown in the top left quadrant. In a half hour when demand is Q, price will be P. A price curve corresponding to the load curve can be plotted in the bottom right quadrant by reflecting off the 45 percent line as shown. Price exceeds P in H half hours of the year. Figure 3.2 uses this price curve to assess the annual profit of a power station with marginal cost c. The station will run for J half hours, and its profit will be the shaded area. Over time there will be an incentive to invest in plant for which this profit, discounted over time, exceeds fixed cost. There will be incentives for the industry to invest to achieve the optimal scale and mix of plant.

In this model there is no "capacity element." How does this enter the

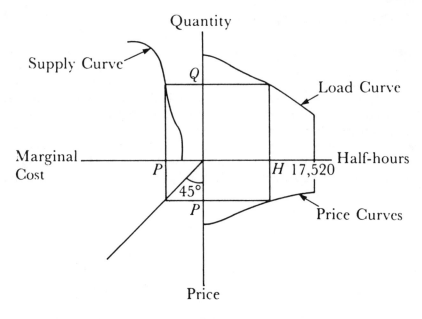

Figure 3.1.   Electricity Demand and Price

picture? The answer has to do with rationing, which may occur because of supply or demand shocks that happen too fast for price responses to equilibrate supply and demand. When rationing occurs, consumer surplus is lost (this is the idea behind the value of lost load). For periods when this might happen, it is appropriate to add an element to reflect the probability of lost load. This encourages the building of capacity and somewhat constrains demand, thereby reducing the congestion externality. Note that the capacity element is set by regulatory intervention, so the market is no longer completely unregulated.

Leaving aside these issues to do with rationing, let us now turn to the issue of market power on the supply side. Even the simplest oligopoly models suggest that this might be a serious concern in the circumstances of the ESI. Because prices rather than quantities are offered to the pool, Bertrand competition is a more natural initial approach than Cournot competition. With constant marginal cost level c and large enough capacities, the unique Bertrand equilibrium indeed has price equal to c. For example, if total capacity K is divided equally between n firms, and if demand when price equals c is $Q(c)$, then marginal cost pricing is the unique equilibrium if and only if $n > K/[K - Q(c)]$. In other words, capacity must be divided

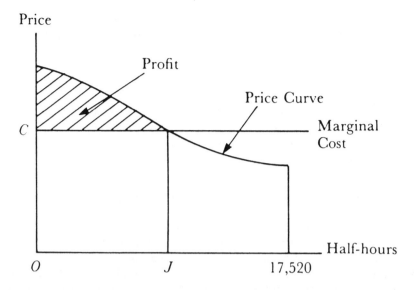

Figure 3.2. The Profitability of a Price-Taking Plant

between a large enough number of firms for capacity constraints not to upset the marginal-cost-pricing equilibrium.

This kind of condition plainly does not hold in the British ESI in the majority of demand states. The analytical difficulty when capacity constraints bind is that equilibrium in pure strategies fails to exist (the way that demand is rationed between firms also needs to be specified). Mixed-strategy equilibrium must be analyzed (see Kreps and Scheinkman, 1983; Tirole, 1988). The general conclusion is that high price-cost markups emerge from this one-shot model when capacity and demand parameters are set to correspond with the conditions that hold for most of the time in the ESI. Except in low-demand states it would indeed be a dominated strategy in this model for a firm in a position like that of National Power to offer a price anywhere near the level of marginal cost. Even if it supplied at full capacity, the profit would be less than that achieved by a high-price strategy of monopolizing the residual demand left after the other firms have supplied at their capacities.

Note that these static models predict high margins even though behavior is entirely noncollusive. In practice firms are in a dynamic relationship, with daily setting of (half hourly) prices. In repeated game models, tacit collusion can be sustained by threats to revert to noncooperative behavior (see Tirole, 1988, for an account of the main results). The degree of such collusion may be limited if the number of firms is large, if discounting is high, or if detection lags are long, but none of these things is remotely true of the British ESI, where, in terms of the models, dream conditions for collusion prevail.

Given that these models suggest that the prospects for competitive outcomes in generation are gloomy, it may be asked whether there exists any defense, other than (the threat of) direct regulatory intervention in the event of excessive prices, against the exploitation of market power. Several factors, absent from the models looked at so far, need to be examined.

3.3.2. Contracts, Entry, and Regulatory Intervention

The first is *contracts* (see Anderson, 1990). Suppose National Power was committed, via long-term contracts struck in the past, to supply a designated number of units of output at a designated price. Then the economic profit on the contract in any particular period would be equal to the specified output multiplied by the difference between the contract price and the pool price (the latter being the opportunity cost of supply). Since any move by National Power to raise prices in pool would depress the profit from the long-term contract, the existence of such a contract serves to reduce the incentives to behave restrictively.

When the new industry structure was introduced in March 1990, the government imposed an initial set of contracts (called *contracts for differences*) between the RECs and the generators—the most substantial of which are of three years duration—which do indeed reduce the incentives of National Power and PowerGen to raise pool prices. The fact that such contracts tend to diminish market power suggests that, once the initial government-imposed contracts run out, the generators might seek to limit the number of contracts that they sell. There are, however, pressures in the other direction, which, in addition to the factors discussed in Sections 3.2.1 and 3.2.2, could include the strategic motive of selling contracts to induce output contractions from rivals. And, unlike the daily repetitions of the spot-market game, conditions for collusion in the contract market are distinctly less favorable: longer contracts reduce the relevant discount factor, and secret price cuts are much harder to observe. The relative ease of collusion therefore suggests that spot prices might be higher than contract prices.

The second key factor is *entry*, which is considered in more detail in the next section. Import competition from Scotland and France is the main immediate threat, but it is limited by the capacity of the interconnectors. Given the time lags in constructing new plant, entrants will take a while

Table 3.1. Pool Selling Prices: Comparison with the Bulk Supply Tariff (BST) (in Pounds per MWh)

	Lowest price in the day	Highest price in the day
Pool (August 22, 1990)	£11.31	£24.53
Pool (August 29, 1990)	11.35	23.86
BST, summer rates (1988–1989)	14.00	26.20

Source: Daily Telegraph; Handbook of Electricity Statistics (1989).

to establish themselves at a substantial scale, though the period required for the introduction of new combined-cycle gas-turbine (CCGT) plant is not long by past industry standards. Thus a purchaser facing high spot market prices because of opportunistic behavior by incumbent generators could enter a long-term contract with a new producer (or build its own plant) for supplies beginning, say, four years hence. That would threaten the generator with stranded capacity in the future. Put another way, the ability to strike long-term contracts gives new entrants the capacity to make preemptive (pricing) moves against incumbents and may therefore serve to increase the competitive pressures that potential competitors exert on established firms. Purchasers, notably RECs, might also contract for new supplies to reduce the market power of the incumbent generators. However, free-rider problems between purchasers could limit this, and it does not deter the generators from exercising their market power in the short term anyway. What might do the latter is fear on the part of incumbent generators that a reputation for short-term opportunism would jeopardize longer-term business. The strength of such a reputation mechanism is a matter for debate.

The third factor is the fear that the exploitation of market power in the short term would *trigger regulatory intervention*. This would be the worst of both worlds for the incumbent generators: the longer-term reputation loss would have been incurred for minimal short-term gain. (It would also be rather embarrassing for the government's reforms, which are based on the idea of competition replacing regulation in generation.) The incumbents must strike a balance between short-term profit and the desire to avoid this risk. If the fear of regulation is an important curb on the exercise of market power, then there is a sense in which the wholesale market is subject to regulation after all. At any rate, it is not characterized by laisser-faire competition as normally understood.

There is some evidence about pricing behavior from the first few months of the workings of the system (see Table 3.1). The evidence is limited, and it must be treated cautiously because the period in the run-up to privatization may not be representative of what happens later. For example, all sorts of strategic signaling might be going on between participants in the industry. That said, initial prices in the preprivatization period do not indicate price-cost margins at the very high levels that are implied by duopoly models without contracts, entry threats, and the fear of regulatory intervention, although there is some indication that the dispersion of time-of-day prices has increased. However, whether or not the generators behave strategically with respect to the capacity element in the pool price—for example, by underdeclaring available capacity in order to raise the loss of load probability—remains to be seen (this is likely to be more of a problem in the winter).

Another important piece of evidence comes from investors' valuations of the incumbent generators. Even allowing for the incentive of the financial community to talk down the offer prices, it does appear very unlikely that the government will be able to sell either National Power or PowerGen at a price even close to their balance sheet (current cost accounting) asset values. For what it is worth, this evidence suggests that the capital market does not perceive a large pot of monopoly profits on the generation side of the electricity industry in the long run.

3.3.3. Oligopoly: Conclusions

The analysis of oligopoly is beset with many difficulties even in the simplest of settings. In the electricity supply industry further complications arise from issues of capacity constraints, the mix of plant, the nature of dynamic interaction, contracts, potential competition, and the threat of regulatory intervention. Given that conditions for collusion between the incumbent generators appear on the surface to be so favorable, at least in the short term, we believe this last factor to be very important. But it implies that the wholesale market is not really free from regulation. Rather, it suggests that a kind of limit pricing may operate, where the danger is entry by the regulator. Although price regulation is usually associated with monopoly or dominant firms, the regulation of oligopoly is not unprecedented: the U.K. salt duopoly is subject to price controls, for example (see Monopolies and Merger Commission, 1986). It is therefore conceivable that the question of wholesale electricity prices might eventually be referred to the Monopolies and Merger Commission (MMC) by the DGES, but whether or not this happens, the possibility of regulatory intervention is likely to be a constraining influence on the exercise of market power in the new electricity supply industry.

3.4. Entry and Access

Given the question marks hanging over the effectiveness of competition among incumbent firms in the wholesale power market, it remains to be considered whether the reform of the ESI can be expected to lead to a substantial increase in competitive pressures as a result of lowering of entry barriers. Since the difficulties of entering the transportation activities (transmission and distribution) are very substantial because of natural monopoly conditions combined with sunk costs, we focus on generation and supply. We start with generation because this is the area where there is the greater scope for reducing costs and because the most radical of the structural reforms (vertical separation of generation and transmission) was targeted chiefly on promoting competition in generation.

3.4.1. Entry into Generation

At present the industry has excess capacity overall, but an inefficient mix of plant. The nuclear program is effectively at a standstill due to high real interest rates and following revelations about costs of waste disposal and decommissioning. Environmental measures (see Section 3.6) are likely to raise the relative cost of coal-fired stations. Technological advance has made combined-cycle gas-turbine plant much more efficient, and in the near-term new generation capacity is therefore likely to be CCGT. Over the longer run the need for new capacity will depend on the rates of demand growth and the retirement of older capacity.

Consider first whether the main incumbent generators, National Power and PowerGen, have intrinsic cost advantages over potential entrants (strategic factors, for example, arising from first-mover advantages, will be examined later). Comparing the incumbents' existing plant with potential new plant, capital costs are to a large degree sunk in the case of the former. However, this is not entirely so. For example, existing coal-fired stations may require expensive investment in flue gas desulphurization (FGD) equipment. Moreover, much of the incumbents' existing plant has relatively high operating costs because of its age and mix. Environmental measures will further increase those costs. In addition, coal prices paid by the ESI have in the past been artificially high. This will continue in the short term, although the implicit subsidies to the domestic coal industry are being steadily reduced.

As for competition to construct new plant, where the track record of the incumbents is poor (see Monopolies and Mergers Commission, 1981), National Power and PowerGen do not have obvious advantages due to

their scale, and their vertical integration with transmission has now ceased. The main potential competitors are themselves large international players—in the oil, gas, power equipment, and construction industries for example—with no capital market disadvantages. Some of them have incentives for vertical integration (see Section 3.2 above)—for example, forward integration by fuel and equipment suppliers and backward integration by RECs and large customers. Autogeneration has a tax advantage in that it is exempt from the fossil-fuel levy, and combined heat and power (CHP) projects are also favored. Without a wholesale market larger firms can provide a given supply security level with proportionately less plant than smaller firms because of the law of large numbers, but in principle the pool should remove any incumbent advantages on this score.

The cost advantages of the incumbents are more related to locational issues, where there are at least two problems. First, they have the great asset of existing sites with planning permission for new plant. Environmental objections are increasing the difficulty and cost of developing new sites (an example of conflict between competition and environmental concerns). Second, the present configuration of the transmission system is closely related to existing power station locations and hence to the sites of the incumbents.

Turning to the possibility of strategic behavior, the first question is whether the incumbents can credibly threaten to cut prices in the event of entry to a level that renders entry unprofitable. Capital intensity, durability, and asset specificity in generation suggest that this might be possible without the incumbents pricing below their variable costs. On the other hand, since plant is of many vintages and types, the variable costs of marginal plant can be high, indeed higher than the total cost of optimal new plant. Selective price cuts targeted against newcomers are impossible given the nature of the market, and sunk costs imply that it would be hard to make an entrant withdraw from the market once it had come in, though aggressiveness might still have reputational advantages for incumbents.

The next issue is whether incumbents will rationally preempt entry by building capacity ahead of rivals, rather as in the preemptive patenting literature (Gilbert and Newbery, 1982). The idea here is that competition will be greater, and hence industry profits will be lower, if a newcomer rather than an incumbent builds the next unit of capacity. With a monopolist or perfectly colluding incumbents, it follows that the incumbent has a greater incentive to build the next unit, but this argument does not extend to noncolluding oligopolists because there are free-rider problems in entry deterrence between incumbents (Gilbert and Vives, 1986; Vickers, 1985).

Virtually all new power projects are for CCGT plant, thanks to the

Table 3.2. New Power Projects Announced by October 1990

Location	Size/type	Lead company
Killingholme	900 MW CCGT	National Power
Rye House	680 MW CCGT	National Power
Killingholme	1000 MW CCGT	PowerGen
Little Barford	680 MW CCGT	PowerGen
Roosecote	220 MW CCGT	ABB
Wilton	1750 MW CCGT	Enron

Source: Power in Europe (1990).

changes in the economics of coal versus gas (see Table 3.2). There are indications that the incumbents, National Power and PowerGen, have been moving more quickly than most competitors into new CCGT construction. However, it is equally evident that there is a great deal of interest in entry, unlike in the period after the 1983 Energy Act. Firms seeking backward or forward integration—fuel suppliers, power equipment makers, construction companies, distribution companies—are particularly prominent. Imports from Scotland also appear to be expanding.

Finally it can be asked whether entry, or the threat of it, may lead to excess capacity in the industry. This could happen for two reasons. The first is preemptive capacity building by incumbents, which has just been discussed. The second is that oligopolistic collusion on pricing, the subject of Section 3.3 below, could lead to high profits that are competed away by capacity building, with the inefficient result of over-capacity in the industry.

3.4.2. Entry into Supply

The entry problem in supply—the acquisition of electricity and its sale to final consumers—stems from the supplier's dependence on incumbent competitors. In a few cases suppliers can bypass the RECs by taking lines straight from the grid, but generally they must use REC wires. One response to the possibility that the regional monopolies in distribution will extend to supply would be complete separation: just as transmission was separated from generation, so distribution could have been separated from supply. However, that would entail the loss of economies of scope between the two activities, which, particularly for smaller loads, may be significant. An alternative approach, and the one that has been followed, is accounting separation accompanied by regulation: the RECs must keep separate accounts for separate businesses, and their distribution charges are regulated. There remain problems of monitoring (such as the fine details of connection arrangements) to ensure that the RECs treat their own and their rivals' supply businesses equally. On the other hand, RECs might not go to great lengths to hold on to their initial dominant positions in supply, especially if competition squeezes margins, and may be content to focus on profit from their wires businesses (recall that, in terms of value added, distribution is a much more significant activity than supply).

In fact, public policy has sought to restrict the competition faced by RECs in supply, by means of the franchises and quotas limiting National Power and PowerGen described in Section 3.3 above. As in telecommunications, where British Telecom and Mercury were protected from competition in fixed-link network operation, dominant incumbents have been shielded from competition in electricity supply. Such limitations on competition may enhance sales proceeds and may be attractive to the protected incumbent firms, but they are questionable on efficiency grounds.

The integration of generation and supply is a factor that might assist entry into supply by generators but hinder competition from others, if generators are able to operate a vertical price squeeze. Such a strategy involves raising the wholesale price of electricity on the spot market, thereby increasing the costs of competitors in supply, and undercutting them in the retail market. Again, accounting separation and regulation against cross-subsidy are intended to cope with the problem, and the alternative policy of rigidly separating generation and supply would have removed an important source of competition to RECs in supply.

Indeed, National Power and PowerGen achieved a very rapid rate of market penetration at the large-load (chiefly industrial) end of the market once competition in supply was opened up. By the end of May 1990, just two months after the new regime came into being, they had captured 9 percent of the market formerly held by the RECs. Scottish Hydro also won a contract to supply twelve sites of BOC, one of the United Kingdom's largest industrial consumers with a maximum demand of about 250 MW. Hydro has also contracted to supply RECs directly. Its strategy is especially interesting because its share of capacity on the interconnector between Scotland and the English and Welsh system is limited, so that it will have to buy from the pool to meet its commitments. Nevertheless its entry into supply has not been deterred.

More generally, Table 3.3 shows the market shares captured by second-tier suppliers—including RECs supplying outside their own areas, as well as companies like National Power, PowerGen, and Scottish Hydro—in the first three months of the new regime. These data make it clear that significant competition in the supply of electricity is already a reality.

Table 3.3.   Percentage Market Shares of Second-Tier Suppliers, July 1, 1990

Region	Nonfranchised market	Total market
Eastern	16.0%	3.5%
East Midlands	19.2	5.7
London	39.0	7.0
Manweb	49.0	25.0
Midlands	33.0	9.9
Northern	57.0	26.0
Norweb	36.7	10.9
South East	44.4	8.7
Southern	32.2	6.7
South Wales	71.0	36.0
South West	62.6	13.6
Yorkshire	42.5	18.7

Source: RECs' privatization prospectuses.

3.4.3. Entry and Access: Conclusions

In the short time since regulatory reform took place in the U.K. electricity supply industry, there has been a good deal of evidence of new entry in both generation and supply. This contrasts with the disappointing results of liberalizing generation in the 1983 Energy Act, and it is interesting to ask why this is so. Part of the reason is that technological advances, factor price changes, and environmental policy have made CCGT plant very attractive, and entrants as well as incumbents have been taking that investment opportunity. However, the reformed regulatory framework has been an essential condition in our view. The weakness of the 1983 Act was the lack of regulation for competition, in particular regulation of access terms. That deficiency has now been made good, and it is the effects of (actual and threatened) new entry on plant construction costs and fuel input costs that are likely to represent the most significant source of benefits from the recent regulatory reforms.

3.5. Price Regulation

As already noted, the scope for increasing competition (actual or potential) in the naturally monopolistic activities of transmission and distribution is highly limited. In respect of these activities, therefore, the emphasis of the policy reforms accompanying privatization has been on the development of new methods of controlling prices so as to prevent abuse of monopoly power.

RPI-X regulation applies separately to transmission, distribution, and supply to smaller consumers. The general merits of RPI-X regulation and its relation to rate-of-return regulation have been much discussed elsewhere: see, for example, Vickers and Yarrow (1988) and the symposium on price-cap regulation in the *Rand Journal of Economics* (1989). Here we focus on both the specific implications of RPI-X regulation for the electricity market and the issues that regulation of the British electricity supply industry has highlighted but that are potentially also of importance to other industries.

3.5.1. Transmission charges

The analysis of price regulation usually assumes that the firm maximizes profit subject to regulatory constraint. In the case of the National Grid Company, however, a distinction must be made between the profit from its own activities and the profits of its owners, the twelve RECs. Because of NGC's key position in the market and its access to commercially sensitive information from all major participants, there are restrictions on the extent to which its owners can control its behavior. Moreover, the transmission license gives NGC some quasi-regulatory duties, such as the duty to facilitate competition, which will sometimes conflict with the RECs' objectives. However, we begin by supposing that NGC's objective is to maximize profit from its transmission activities, as if it were an independent entity.

Because of its high degree of both spatial and temporal differentiation, electricity transmission provides some graphic illustrations of the difficulties of price regulation in multiproduct industries. For example, should regulators focus on constraining some relatively simple overall price index, leaving detailed aspects of the price structure to be determined by the regulated firm itself, or should regulators attempt to control the structure as well as the overall level of prices? The first option risks the introduction of pricing distortions as the firm seeks to minimize the effect of the regulatory constraint on profits, whereas the latter runs the risk of distortions due to flawed regulatory decisions as a result, for example, of limited information.

The initial tariff structure for the privatized National Grid in England and Wales represents an ad hoc mixture of different approaches (Table 3.4 gives an outline). The main distinction is between charges for connection to the grid and for the use of the system. The former are intended to cover the costs (mostly capital costs) incurred in providing points of entry, where

Table 3.4. Initial Tariff Structure for the National Grid (in Pounds)

Average annual entry charge	£1.25/kW
Average annual exit charge	£4.00/kW
Proposed system service charge	£3.37/kW
Suppliers' infrastructure charges (zonal)	£5.96–8.50/kW
Generators' infrastructure charges (capacity)	£0.00–3.10/kW
Generators' infrastructure charges (energy)	£0.00025/kWh

Source: Holmes (1990), from NGC.

generators connect to the grid, and points of exit, where RECs and other electricity suppliers draw power from the grid. Entry and exit capital is frequently specific to one or a small number of generators or suppliers. Thus, while entry and exit charges for existing facilities are included within the overall price cap for transmission, connection charges for users requiring additional facilities are excluded from the cap and are separately regulated on a rate-of-return basis. The exclusion can be seen as an attempt to prevent price-cap regulation from discouraging new investment in access facilities.

Use-of-system charges are regulated according to an RPI-X formula (with X set equal to zero) that specifies an upper limit on NGC's revenue per unit of (average cold spell) maximum demand on the system. Use-of-system charges include a per kW system charge levied on all suppliers, which is claimed to reflect the benefit that all suppliers derive from the existence of the grid system; generators' infrastructure charges, levied on both power (per kW) and energy (per kWh), with variations in charge levels among eleven regional zones; and per kW suppliers' infrastructure charges, also zonally differentiated.

Since volume for the purposes of the pricing formula is defined as maximum demand on the system, profit-maximizing behavior implies NGC seeking to increase that maximum demand at least cost. This means encouraging location patterns of incremental demand and supply that do not require major new investments in reinforcing the system. While it may be argued that this will encourage optimum use of existing transmission assets, when the multiproduct nature of transmission services and their implications for generation are taken into account, the incentive system could go too far in this direction. Consider, for example, two power station projects of equal size, located at different points of the grid, one in the north and one in the south. Suppose that the north is a net exporter of power, that the north to south flow is limited by transmission constraints, and that the incremental demand giving rise to the requirement for new power projects lies in the south. Finally, let the incremental generation and transmission costs be G_n, G_s, T_n, and T_s, where $T_n > T_s$ and $G_n + T_n < G_s + T_s$. That is, incremental transmission costs are higher for the power station in the north, but the overall cost of delivered power is lower, and so efficiency is higher, if the northern location is chosen.

Now a profit-maximizing transmission company subject to RPI-X regulation of the British type will prefer the less efficient project. That is because the revenue implications of the two projects are the same, but the transmission costs associated with locating the power station in the south are lower. The grid company might therefore seek to discourage new generation investment in the north directly, by quantitatively limiting access to the system, or indirectly, via a charging structure with similar effect. The general point is that there is a bias—resulting from a vertical externality and pushing toward regional autarky—built into the type of price control formula under discussion. As well as potentially leading toward short-run resource misallocation, this could clearly also have adverse long-term effects on the state of competition in electricity generation.

Thus far the analysis has been based on the assumption that the transmission company maximizes profits, but it might be argued that suboptimal project selection would raise the costs of RECs and that the latter would therefore have every reason to use their influence as owners of the grid to discourage the distorted pricing structure. The strength of the point is weakened, however, by the facts that (1) the value to RECs of cost savings is limited by the cost-pass-through provisions of the pricing formula for the franchise market, (2) there are twelve RECs and there may be conflicts of interest among them in respect of particular grid reinforcement investments, and (3) as well as facilitating additional competition in generation, grid reinforcement might increase competition in supply, to the detriment of the RECs' own supply businesses. For example, from the perspective of RECs, the construction of additional transmission links to France might be viewed as a two-edged sword: on the one hand it might increase competition in generation, to the benefit of RECs; on the other hand EdF might win business from RECs in supplying the nonfranchised electricity markets.

In our view, the fact that is most likely to modify the distorted incentives implicit in the RPI-X formula is the prospect of regulatory intervention. Among the more obvious limitations of the existing structure are the following: (1) there is no tariff component that reflects the costs of transmission losses (see Section 3.2.2 above), (2) the degree of spatial differentiation in charges appears to be significantly less that the corresponding differentiation in costs, (3) the system service charge is not closely linked to an identifiable category of costs, and (4) there is no temporal disaggregation of charges (for example, a generator with given capacity would pay the same transmission charges whatever the time profile of its supply).

Unsurprisingly then, the Director General of Electricity Supply has already indicated his desire to see changes made, and his first annual report outlined a timetable that sees revised transmission charges being phased in from April 1993 onward. The initial pricing structure for the grid must therefore be regarded as something of a holding operation. It is likely to be changed in the not too distant future; but how it will be changed is unclear. The regulatory uncertainty that exists in the meantime complicates investment planning—especially location decisions—in both generation and transmission.

3.5.2. Distribution and Supply

An important innovation of the electricity privatization exercise was the distinction made, in PES licenses, between the distribution and supply businesses of the RECs. The two businesses are subject to differing price-cap formulas, and the supply business of any REC must purchase electricity transportation services from its own distribution business on terms similar to those available to a competing electricity supplier. This approach—which aims for accounting and business separation of naturally monopolistic distribution from potentially competitive supply—is in marked contrast to the regulatory decisions taken at the time of the gas privatization in 1986 (when accounting separation between the carriage and supply of gas was not imposed). Because it will facilitate public monitoring of the terms on which competing suppliers will have access to the wires of the RECs, the innovation represents a significant improvement in the implementation of regulatory policy.

Since electricity distribution is similar to transmission in that it is a transport activity, many of the issues raised by the regulation of distribution charges are similar to those surrounding the regulation of transmission rates (and the criticisms cited in Section 3.5.1 also apply to the structure of distribution charges). We therefore focus here on questions concerning the regulation of supply charges to smaller customers.

In respect of supply, a key point to note is that the value added contributed by REC electricity supply businesses amounts to something of the order of only 4 percent of the final selling price. The remainder is accounted for by the RECs' purchase costs, which comprise the purchase costs of bulk electricity, the fossil-fuel levy, transmission charges, and distribution charges. To a large (but not total) extent, these purchase costs are beyond the control of a REC's supply business. Thus, for example, when there is a major hike in fossil-fuel prices, RECs will find it almost impossible to avoid large increases in their purchase costs.

In the above circumstances, a substantial degree of pass-through of purchase costs can be justified. Linking price changes to (marginal) cost changes is generally good for allocative efficiency, and the loss of cost-reduction incentives will be limited if the relevant cost components are genuinely beyond the control of the firm. Where, however, the regulated firm has more influence on the relevant cost items, it will tend to be more efficient to allow only partial indexation of prices: allocative efficiency is sacrificed to improve overall efficiency by strengthening incentives for cost reduction.

In England and Wales each REC will be allowed to pass through 100 percent of changes in its own bulk electricity purchase costs, the fossil-fuel levy, distribution charges, and transmission charges (all of which are included in the Y-factor in the pricing formula). However, this was not always the intention of the government. In the first published versions of the draft PES licenses in early 1989, pass-through of bulk electricity purchase costs was linked to a yardstick formula that implied that pass-through would be greater (less) than 100 percent if an individual REC was more (less) successful than other RECs in restraining its bulk purchase costs. In this way, it was intended that RECs would face strong incentives to strive for lower bulk electricity prices and thereby increase the pressures on generators to improve upstream efficiency.

In principle, yardstick regulation is a potentially effective way of making use of relevant economic information to improve the tradeoff between allocative and cost efficiency. However, the attempt to turn general principle into regulatory practice was not particularly well thought through. It was never clear, for example, why only bulk electricity purchase costs were singled out for the yardstick treatment when cost components over which RECs had more influence were not. The yardstick proposal was also linked to plans envisaging a relatively complex contractual structure linking suppliers and generators. These plans were later abandoned.

It can be argued that the simplicity of the final arrangements for purchasing bulk electricity, and in particular the availability of supplies from the pool, means that there is little ground for concern about the incentives effects of 100 percent cost pass-through, particularly as RECs come under increasing competitive pressures in their supply businesses. The blunting of incentives to reduce costs by the application of effort—the effect analyzed in many applications of principal-agent theory to the regulatory problem—

is not, however, the only negative consequence of the cost pass-through provisions. Difficulties also arise as a result of interactions between price regulation and vertical integration in the industry, which in some circumstances might provide RECs with incentives to raise costs.

Recall that each REC owns a stake in the National Grid Company and, under the terms of its license, is allowed to generate up to 15 percent of its own electricity requirements. As explained in Section 3.5.1 above, it is possible to envision circumstances where RECs would prefer higher generation plus transmission costs because (1) the profits of transmission activities are increased as a result and (2) the higher overall costs can be passed forward to consumers. Similarly, a REC could find it beneficial to have higher bulk-purchase prices if it can participate in the higher generation profits that this implies (see Helm and Yarrow, 1988). The general point here is that vertical integration can hinder effective regulation by making it more difficult to prevent adverse spillover effects of price controls.

3.5.3. Price Regulation: Conclusions

The general problem with regulation is that, although its purpose is to redress market failure problems, it can induce distortions of its own—in pricing, cost reduction, and investment decisions. Early theoretical work on U.S. rate-of-return regulation suggested dangers of overinvestment, among other problems. In the United Kingdom, RPI-X price cap regulation was partly intended to overcome some of these perceived problems. However, depending on the information available for regulation, RPI-X might not be very different from rate-of-return regulation (with long lags). Moreover, it faces problems of its own, for instance the risk that policy credibility problems may lead to underinvestment. This is an example of a problem that can occur in the simplest of settings with a single homogeneous good, but we have seen how it can be important also in the complex circumstances of the electricity supply industry.

The importance of the *structure* of regulated prices is illustrated by the question of grid pricing. The average price cap might introduce distortions that in turn affect investment decisions in generation as well as transmission. This is not to say that there are alternative easy answers—quite the reverse is true—but we hope to have indicated that the value of designing more efficient schemes is potentially very great. Above all, the vertically deintegrated electricity supply industry illustrates the important point that price regulation at one stage of the supply chain (such as transmission) can have major effects on competition at another stage (such as generation).

3.6. Environmental Regulation

Perhaps the most important issues facing the electricity supply industry in the coming years have to do with environmental protection. Environmental regulation has been analyzed elsewhere (see Newbery, 1990), and the main concern of the current paper is with the industrial organization of the ESI. However, the issues are closely related, and in this section we examine two specific linkages: (1) possible investment problems induced by environmental regulation (which are particularly acute with nuclear power) and (2) the fossil-fuel levy and its relation to the problem of carbon dioxide emissions and the greenhouse effect.

3.6.1. Investment Issues

Intertemporal considerations affect environmental regulation because of investment problems of the kind already discussed in other contexts. A firm making sunk investments faces the risk that subsequent changes in environmental policy may require additional investments, such as the retrofitting of FGD equipment, which, if anticipated at the outset, would have rendered the project unprofitable. The type of risk is present even if the product markets into which the firm sells are unregulated. However, additional problems arise when, as we have argued is implicitly the case for bulk electricity, product prices are subject to regulatory influence. Thus policy makers may find it difficult to make credible commitments that future expenditures required by a tightening in environmental standards will be reflected in higher allowable product prices. In these circumstances, firms might face the prospect of falling profits as a result of political unwillingness to allow recovery, from consumers, of the costs of environmental protection.

Nuclear generation in the United States provides a good example of the type of effect just described. Although it was not the only factor at work, the additional costs imposed by more stringent environmental regulation in the wake of events such as the Three Mile Island accident revealed the vulnerability of profitability to policy changes. Regulatory commissions were unwilling to allow the escalating costs of nuclear power—to which cost overruns and higher interest rates were also contributory factors—to be passed through to consumers. As a consequence, there was a large-scale abandonment of nuclear projects, including power stations already under construction.

The threat of radiation release, such as occurred at Three Mile Island, is only one of the environmental considerations relevant to investment in nuclear-generation capacity. More important in the United Kingdom has been the problem of disposal of nuclear waste. The safe disposal of such wastes imposes liabilities on generators that are highly uncertain and, more important here, whose value is highly susceptible to future public policy decisions as to what are and are not considered to be acceptable methods of disposal. The result is another major policy credibility problem, which adds to the perceived costs of electricity produced from privately owned nuclear power stations. In the event, initial plans to privatize nuclear generating capacity in the United Kingdom, set out in the 1988 White Paper, were abandoned in 1989, and the future nuclear program (excluding the station currently under construction at Sizewell) has been frozen.

3.6.2. Greenhouse Gases and the Fossil-Fuel Levy

At first sight it may appear ironic that the nuclear power program has been mothballed at a time when increasing concern about global warming is anticipated to lead to environmental measures aimed at limiting atmospheric emissions of carbon dioxide. One of the possible regulatory instruments that might be applied in connection with this problem is a "carbon tax" levied on fossil fuels, which will tend to raise the costs of generating electricity from some of the main technologies competing with nuclear stations, particularly coal.

In fact, as part of the privatization exercise, the government has introduced a fossil-fuel levy that will be applied to electricity generated by coal-, oil-, and gas-fired power stations. The proceeds of this levy will be paid to Nuclear Electric, thereby simultaneously taxing power from fossil-fuel stations and subsidizing nuclear power. The levy originated from a long-standing government commitment to nuclear power that had more to do with a desire to avoid heavy reliance on domestically produced coal—where security of supply has been disrupted by industrial action in the past—than to environmental concerns, although the latter have been used as an ex post justification for the levy. And, although the levy might be viewed as a rough, first approximation to a carbon tax, there are a number of important differences. For example, the levy bears down only on fossil fuel used to generate electricity and its base is the unit of electricity produced, not the amount of carbon dioxide emitted in producing that unit (which is lower for gas than for coal or oil).

That the fossil-fuel levy proved inadequate to make the privatization of nuclear power a viable proposition reflects in part the strength of the policy credibility problem for a technology as capital intensive and politically sensitive (for environmental reasons) as nuclear electricity generation. For example, simply raising the magnitude of the levy would not have disposed of the problem, since it would have been difficult for the government to commit to maintaining the levy over long periods.

Of the principal technologies used to generate electricity in Britain, the one most favored by tightening environmental regulation is CCGT. Policy credibility problems, and the more general uncertainties surrounding the introduction of the new market and regulatory structures, tend to favor CCGT because it is less capital intensive than nuclear, coal, or oil. In addition, the risks posed to investors by changing environmental policies are rather lower than for coal and oil. If, say, a carbon tax were introduced at some point in the future then, because it produces less carbon dioxide per kWh generated, such a tax would increase the cost of CCGT technology by less than the costs of existing coal and oil technologies.

3.6.3. Environmental Regulation: Conclusions

The major policy reforms surrounding the privatization of the ESI in Britain were not developed with a view to tackling environmental problems. Nevertheless, there can be little doubt that environmental regulation will have a strong influence on the future conduct and performance of the industry. Indeed, such regulation is increasingly identified, worldwide, as the most important challenge facing the electricity industry in the 1990s.

The growing priority assigned to environmental protection raises major questions about how the new regime will function alongside environmental regulation. Answers must await evidence on how the experiment works in the post-privatization period. Meanwhile, however, it is safe to conclude that much work remains to be done in understanding the interactions between regulation of market power and regulation for environmental protection, an area of analysis that is still very much in its infancy.

3.7. Concluding Remarks

Solutions to the multiple market failures that occur in electricity supply have traditionally involved high degrees of both vertical and horizontal coordination, whether by common ownership or by collaboration among independent firms. Although one or other version of the traditional approach is still predominant in virtually all countries, there has been increasing interest among policy makers in moving away from integrated structures, at least to some degree. Whereas this process has been cautious and gradual elsewhere, Britain has just taken the radical step from nationalized monopoly to a privatized, vertically deintegrated structure with competition in both wholesale and retail supply.

Policy interest in deintegration is not, however, confined to electricity supply. Competition policy more generally is concerned with just such questions, and one of the major issues that arises in many industries is the effect of vertical relationships on horizontal competition. At EC level, for example, special regulatory provisions exist in relation to vertical contractual arrangements in the supply of beer, of petrol, and of motor vehicles (see Whish, 1989). All three of these industries have also been the subject of recent major U.K. competition policy investigations, but whereas vertical agreements were given a clean bill of health in petrol supply (Monopolies and Mergers Commission, 1990), the MMC recommended that similar agreements be weakened in beer supply (Monopolies and Mergers Commission, 1989).

How then does analysis of the British electricity experiment help us to understand better these difficult issues of markets versus hierarchies? In the first place, it illustrates the point made in Section 3.2 that vertical integration and long-term contracts may themselves be market responses to failures and inefficiencies surrounding spot contracting. Public policy that seeks to restrict these responses therefore risks the reintroduction of the underlying market failures, which take the form of vertical and horizontal externalities. This general risk, common to many markets, is particularly acute in electricity because of economic characteristics of transmission grids (Section 3.2.2).

The main argument *for* vertical separation is that it may promote horizontal competition. In respect of supply of electricity to the nonfranchised market, the partial separation of supply and distribution has indeed led to a surge in competition for the accounts of large end users (Table 3.3). The supply business, however, accounts for only a small fraction of final electricity prices, and it requires little in the way of capital to enter: the much more important issue concerns the effects of the policy reforms on electricity generation. In respect of generation we have argued that, in the absence of regulation, the market conditions created will be conducive to noncompetitive pricing behavior, including tacit collusion (Section 3.3). Implicit regulation of the bulk market is therefore likely to persist.

One possible conclusion from this is that the reforms were not radical enough: they should have gone further in promoting horizontal competition by, for example, splitting the CEGB into a larger number of generating companies. Since the creation of the National Power and PowerGen duopoly was largely dictated by the desire to privatize nuclear power, and since nuclear power was eventually dropped from the privatization program, ex post there is indeed little rationale for the existing market structure. However, given some of the key characteristics of the industry (including investment lags, a highly replicated price game for wholesale supplies to the pool, massive spatial and temporal product differentiation through the network), while more radical proposals—such as splitting the CEGB's nonnuclear generation activities into five companies, each of similar size— might have diminished the severity of the problem of collusion, it is far from clear that they would have been sufficient to create really effective competition in the pool.

Where the reforms are likely to have more beneficial effects on competition is via their effect on entry conditions (Section 3.4). Lower barriers to entry in generation will have two main effects: (1) current arrangements whereby the ESI has been used to protect the domestic coal industry will be undermined, and (2) there will be downward pressure on power station construction costs, where past industry performance has been poor.

The importance of entry conditions for competition is a message that has long been stressed in the industrial organization literature, but, accepting the general policy goal of reducing entry barriers, it is not clear that the objective is being achieved in the most efficient way. Protection of the domestic coal industry could have been simply and easily ended by, for example, liberalizing trade in coal. Bidding schemes for new power station construction offer one of several alternative approaches to reducing capital costs by promoting competition. Evidence in favor of the British "structural" approach is the contrast between the entry that is occurring now and the absence of new entrants following the Energy Act 1983. On the other hand, the unfavorable comparisons between past U.K. power station construction costs and those elsewhere in the world might be interpreted as showing that much greater cost efficiency can be achieved without radical restructuring: as already noted, the countries with costs lower than the United Kingdom continue to rely on integration and collaboration in electricity supply. This latter view would suggest that the British electricity experiment may be a case of taking a sledgehammer to crack a nut.

Regulatory reform has not, of course, been directed exclusively at attempting to increase competition: new arrangements for regulating the naturally monopolistic transmission and distribution networks have been put in place. In these areas, the general lessons to be learned concern the relative effectiveness of regulated private monopoly and public monopoly, an issue that has arisen throughout the course of the U.K. privatization program, and one that is relevant to utility industries everywhere.

We have stressed the regulatory complexities that arise from the spatial and vertical characteristics of electric networks (Section 3.5). These characteristics, and the resulting regulatory problems, are, like the monopoly positions themselves, not fundamentally changed by privatization. Initial attempts to deal with many of the problems of price regulation have an explicitly interim aspect to them. However, the laudable, longer-term intention to improve on relatively crude initial tariff structures necessarily comes into conflict with the objective of providing stable incentive structures. This is another problem of vertical relationships: the regulator is like a monopsonist with unilateral powers to vary price (see Section 3.2.1), and one of the dangers of this arrangement is inefficiency in investment. The problems can be countered by either public ownership—a form of vertical integration between regulator and firm—or by a long-term regulatory contract or bargain (implicit or explicit), both of which are to be found elsewhere in the world. Britain has abandoned the first solution but, in effect, has not yet fully embraced the latter. The specter of policy instability, which proved so damaging to the ancien regime, still lurks.

Finally, we have addressed some questions surrounding the environmental regulation of the ESI (Section 3.6). Once more, there is a general, vertical dimension to the problems: the environmental regulator can unilaterally determine the terms on which firms will have access to environmental services, an important input so far as electricity generators are concerned. Familiar policy credibility and underinvestment issues therefore arise, with perhaps the best example being the problems faced in attempting to reconcile nuclear power programs with private ownership.

As a result of increasing public policy concern about the effects of atmospheric emissions of waste gases, and as in many other industries, this particular contractual boundary between the ESI and its regulators is likely to become increasingly important over time. Indeed, environmental regulation can be expected to be *the* major issue facing the ESI, worldwide, in the 1990s. Since the new regulatory framework in Britain was not developed with environmental problems in mind, there is a danger that, at the international level, it will come to be treated as a mere sideshow to the main (environmental) event. If so, that would be a pity; for, as we hope we have shown, the information the experiment promises to yield will be relevant in many contexts, not least in the context of environmental regulation itself. The reforms may not be widely copied, but they do merit close scrutiny.

3.8. The British Electricity Experiment: Postscript

The three years following the initial flotations of electricity companies have witnessed a number of major changes in the market, particularly in generation and supply where privatization was accompanied by measures to increase competition.

The determination of pool prices has proved a particulary contentious issue. There were a number of complaints from industrial users of electricity about the way in which the way in which the pool was operating in the immediate post-privatization period. Complainants were able to point to a number of occasions on which one or more elements of the pool output price (such as the system marginal price, the capacity element, or uplift) moved in apparently perverse ways (such as the system marginal price increasing as demand fell). There was also an overall upward drift in pool prices relative to the price of the primary input (coal).

The concerns of industrial consumers were linked to the fact that, over time, the pool has acquired greater significance in the processes by which the prices paid for electricity by final customers are determined. In 1993, for example, the Director General of Electricity Supply estimated that approximately 1,000 customers (generally of large size) were taking supplies on terms directly related to pool prices, rather than on fixed-price contracts.

A series of investigations of pool pricing movements by the Director General confirmed the obvious: that large generators have the market power to influence prices via their bidding behavior. As a result of the investigations a number of changes were made to the technical details of the bidding rules—for example, concerning the way in which plant availability is declared—with a view to reducing the susceptibility of pool prices to manipulation. Nevertheless, market power remained, and the Director General had to decide whether to refer the two main generators to the Monopolies and Mergers Commission. In February 1994 he announced that having secured undertakings from National Power and PowerGen, he would not make such a reference. The generators agreed (1) to sell or dispose of some capacity, and (2) to bid into the pool in such a way that the average pool price does not exceed a price cap. Insufficient competition in generation has therefore led to the introduction of some price controls after all.

In respect of economic indicators other than pool prices, there have been definite signs of increasing competitive pressures in electricity generation. For example, the combined market share of National Power and PowerGen fell from nearly 78 percent of the electricity sold into the pool

Table 3.5. Market Shares of Consumption in the Nonfranchise Market (in Percentage)

	1990–1991	1992–1993
Local REC (first tier)	62%	51%
National Power	20	12
PowerGen	11	18
Other RECs (second tier)	4	13
Others	3	6

Source: OFFER.

in 1989–1990 to just over 68 percent in 1992–1993. Most of the swing was accounted for by increased output from the state-owned Nuclear Electric, which, under the commercial pressures induced by the new market environment, substantially improved the operating capabilities of its existing power stations. By 1993, however, new entrants were also beginning to supply power on a significant scale.

Both National Power and PowerGen have proceeded with major closure programs of old, coal-fired plant, which the companies will only partially replace with new CCGT stations. Thus, the market shares of the leading generators are projected to continue to fall as more rivals enter the market with their own CCGT plant (eleven such stations are expected to be supplying power by the end of 1995). Partly as a result of plant closures and partly through efficiency gains achieved elsewhere, the numbers employed by National Power and PowerGen roughly halved between 1990 and 1993.

In the retail supply of electricity, the early trend of regional electricity companies to lose market share in the nonfranchise market in their "home" areas has continued. However, as Table 3.5 shows, RECs have been active in competing for business as second-tier suppliers outside their home areas.

Prices in the nonfranchise market fell in real terms in the immediate post-privatization period, although very large electricity consumers did less well than most as a consequence of the withdrawal of special pricing schemes that had existed in the days of public ownership. However, average prices in early 1993 were somewhat higher than would have been expected on the basis of extrapolations of pre-privatization relationships between electricity prices and coal prices. In the franchise market prices actually increased in real terms between 1990 and the first quarter of 1993, and hence the divergence between the trends in final selling prices and in primary fuel costs was even more marked.

Overall, then, it appears that the initial effect of privatization and the accompanying regulatory reforms has been to increase prices, with the larger impacts tending to occur in those parts of the market not open to competition. In this context it can be noted that, despite a sharp recession in the U.K. economy, the profits in the industry increased very rapidly in the early 1990s.

The immediate post-privatization price history of the industry was, however, largely predetermined by the various contracts and price-control formulas put in place in 1990. The initial coal contracts expired in March 1993, causing a major crisis in the coal industry that triggered limited government intervention to try to induce generators again to sign contracts favorable to British Coal. In any event, however, the new contracts provided for further sharp reductions in real coal prices and a substantial contraction in domestic coal volumes. Generation costs fell significantly as a consequence. Moreover, a reduction in the fossil-fuel levy from 11 to 10 percent also contributed to downward cost pressures in the industry.

A review of transmission pricing by the Director General in 1992 led to a tightening of the price controls on the National Grid Company, with the X factor in the RPI-X formula being increased from 0 to 3 percent as from April 1, 1993. Similarly, a review of supply prices in 1993 led to increases in the X factor from 0 to 2 percent. It is expected that the 1994 review of distribution charges will also lead to the imposition of more stringent price caps.

There are, therefore, reasons to believe that the pricing performance of the industry may start to show improvements from 1994 onward. The tighter price controls will benefit franchise market customers directly, but it remains to be seen precisely how customers in the nonfranchise market will fare.

References

Aghion, P., and P. Bolton. 1987. "Entry Prevention through Contracts with Customers." *American Economic Review*, 77, 388–401.

Anderson, R.W. 1990. "Futures Trading for Imperfect Cash Markets: A Survey." In L. Phlips (ed.), *Commodity Futures and Financial Markets*. Amsterdam: Kluwer.

Bohn, R.E., M.C. Caramanis, and F.C. Schweppe. 1984. "Optimal Pricing in Electrical Networks over Space and Time." *Rand Journal of Economics*, 15, 360–76.

Department of Energy. 1988. *Electricity Privatisation*. London: HMSO.

Gilbert, R.J., and D.M. Newbery. 1982. "Preemptive Patenting and the Persistence of Monoploy," *American Economic Review*, 72, 514–26.

Gilbert, R.J., and D.M. Newbery. 1988. "Regulation Games," CEPR Discussion Paper No. 267.

Gilbert, R.J., and X. Vives. 1986. "Entry Deterrence and the Free Rider Problem." *Review of Economic Studies*, 53, 71–83.

Helm, D.R., and G.K. Yarrow. 1988. "The Regulation of Utilities." *Oxford Review of Economic Policy*, 4(2), i–xxxi.

Holmes, A. 1990. *Electricity in Europe*. London: FT Business Information.

Joskow, P.L. 1987. "Contract Duration and Relation-Specific Investments: The Case of Coal." *American Economic Review*, 77, 168–85.

Joskow, P.L. 1989. "Regulatory Failure, Regulatory Reform, and Structural Change in the Electrical Power Industry." *Brookings Papers: Microeconomics*, 125–99.

Joskow, P.L., and R.L. Schmalensee. 1983. *Markets for Power*. Cambridge, Mass.: MIT Press.

Kreps, D., and J. Scheinkman. 1983. "Quantity Precommitment and Bertrand Competition Yield Cournot Outcomes." *Bell Journal of Economics*, 14, 326–37.

Littlechild, S. 1983. *Regulation of British Telecommunications Profitability*. London: HMSO.

Monopolies and Mergers Commission. 1981. *Central Electricity Generating Board*. London: HMSO.

Monopolies and Mergers Commission. 1986. *White Salt*. London: HMSO.

Monopolies and Mergers Commission. 1989. *The Supply of Beer*. London: HMSO.

Monopolies and Mergers Commission. 1990. *The Supply of Petrol*. London: HMSO.

Newbery, D.M. 1990. "Acid Rain," *Economic Policy*, 11, 299–346.

Power in Europe. 1990. London: Financial Times Business Information.

Rand Journal of Economics. 1989. "Symposium on Price-Cap Regulation."

Tirole, J. 1988. *The Theory of Industrial Organization*. Cambridge, Mass.: MIT Press.

Vickers, J.S. 1985. "Pre-emptive Patenting, Joint Ventures and the Persistence of Oligopoly." *International Journal of Industrial Organization*, 3, 261–73.

Vickers, J.S., and G.K. Yarrow. 1988. *Privatization: An Economic Analysis*. Cambridge, Mass.: MIT Press.

Vickers, J.S., and G.K. Yarrow. 1991. "The British Electricity Experiment." *Economic Policy*, 12, 188–232.

Whish, R. 1989. *Competition Law* (2nd ed.). London: Butterworths.

Wilkinson, M. 1989. "Power Monopolies and the Challenge of the Market: American Theory and British Practice," Kennedy School, Harvard University, Discussion Paper E-89-12.

Williamson, O.E. 1975. *Markets and Hierarchies: Analysis and Antitrust Implications*. New York: Free Press.

4 BRITAIN'S UNREGULATED ELECTRICITY POOL

Richard Green[1]

4.1. Introduction

The restructuring and privatization of the electricity supply industry in England and Wales were probably the most ambitious parts of the whole privatization program. This chapter examines the wholesale market that was set up for electricity generators, the centerpiece of the restructuring. Despite the technical demands of an electricity network, the government believed that it would be possible to establish a competitive market for bulk electricity and that no economic regulation would be needed. The mechanisms that it set up have succeeded in the tasks of coordinating generation with demand and paying for it. This chapter goes beyond that technical success to examine the economic success, or failure, of the market, for there has been great concern over the dominance of two generating companies and allegations that they have been able to push prices significantly above their marginal costs.

The privatization of the electricity supply industry in England and Wales was promised in the 1987 Conservative Manifesto, and the restructuring was outlined in a White Paper in the following February (Department of Energy, 1988). Electricity transmission and distribution are natural monopolies and are regulated as such in the new structure, but this does not extend to generation and supply to consumers. The retail activity of supply, which comprises both metering and billing as well as paying for generation, distribution, and transmission, is being opened up in stages; customers with a maximum demand of more than 1 megawatt (MW) may already choose their supplier, while smaller customers must buy from their local distribution company at a regulated price until 1994 (if their demand exceeds 100 kilowatts (kW)) or 1998.

Generation was liberalized completely from vesting day, March 31, 1990, and the only legal limits on entry are planning controls and the need to acquire a license, which enforces technical standards. The former Central Electricity Generating Board was divided into the National Grid Company, responsible for transmission, and two (and later, three) competing generating companies. The government initially hoped to privatize its nuclear power stations (about one-sixth of the total capacity) as part of a large generating company, National Power, but when information on their costs was released, it became clear that they could not be sold, and so Nuclear Electric, a state-owned company, was created. There was no time to subdivide the conventional companies, and so National Power was given about half of the CEGB's capacity, and its smaller rival, PowerGen, was given about one-third. A large number of independent companies (often linked to the regional electricity companies (RECs)) announced plans to build new stations in the two years following vesting day. Almost all of them were combined-cycle gas-turbine stations, which promised high thermal efficiencies and low capital costs. Some of these projects have been stopped, unable to secure a supply of gas, while 4 gigawatts (GW) are awaiting final planning consent from the government, but 10 GW (including 3 GW owned by National Power and PowerGen) are under construction or in operation. Since the total capacity in England and Wales at present is 60 GW, this represents a substantial addition, and it seems likely that a significant number of older stations will have to be closed prematurely if the industry is to avoid overcapacity.

A special market, the pool, has been created to organize the scheduling, dispatch, and payment of power stations on a daily basis. Each generating set submits a multipart bid, consisting of prices and technical information, and a computer algorithm (GOAL) then produces the schedule that minimizes the cost of meeting demand. The bid from the most expensive station scheduled for normal operation in each half hour is used to calculate the system marginal price, paid for every unit scheduled in that period. An additional capacity element, based on the product of the loss of load probability and the value of lost load, is paid to every station that is available, whether or not it is generating in that half hour. These schedules and prices are drawn up a day ahead of operation, and changes in demand, outages, and reactions to transmission constraints will mean that some stations generate more or less than in the unconstrained schedule. The difference is paid for at each station's own price, and the sum of these payments, together with those for some other services required to keep the system stable, is spread over all the units bought from the pool (except in trough periods) to give a unit charge known as *uplift*. This is added to the system marginal price and the capacity element (which make up the pool input price) to give the pool output price. Electricity suppliers pay this price on their metered demands, scaled up for transmission losses, so that adjusted demand equals generation within each half hour, and payments by suppliers equal payments to generators for each day. The appendix describes the pool in more detail.

The pool prices will tend to be volatile, moving with the level of demand, which depends on the weather, and with the level of capacity available, which is affected by random faults. To provide greater certainty, most electricity sales in the pool are covered by contracts for differences, which pay the difference between the pool price and the contract's strike price for a given amount of electricity at a given time. If the parties to the contract also buy and sell that amount of electricity in the pool, then their net payment will be fixed, whenever the contract comes into operation. A one-way contract is triggered only if the pool price exceeds the strike price and gives suppliers the insurance of a known maximum price for electricity. A two-way contract involves payments from the supplier to the generator[2] if the pool price is below the strike price, so that the net payment for electricity is fixed at the strike price.

Although the entrants are rapidly building new capacity, and Nuclear Electric and utilities in France and Scotland are able to send out substantial quantities of baseload electricity, there has long been concern that the decision to create only two conventional generating companies from the CEGB gave them an unhealthy degree of market power. Bolle (1992), von der Fehr and Harbord (1993), and Green and Newbery (1992) studied the pool in isolation, and showed that prices well above marginal cost were sustainable in equilibrium. Green and Newbery also showed that entry was likely to lead to excess capacity, rather than marginal cost pricing, as the means of reducing profits to normal levels. These studies ignored the effects of contracts for differences on the pool, but it soon became apparent that the contracts did have a significant effect in keeping pool prices down (Helm and Powell, 1992). Further work by Powell (1993), von der Fehr and Harbord (1992), and Green (1992) then showed that the contract market did indeed reduce equilibrium prices, although they would still be above marginal cost, in general. Following complaints about the level of pool prices, the regulator published a report (OFFER, 1991) that concluded that, because of the influence of contracts, it was impossible to tell whether pool prices were too high or too low but promised to keep the generators' behavior under consideration.

This chapter tests those theoretical fears against data for some of the generators' bids during the first two years of the new system. The next section describes part of the model of Green (1992), to show the patterns that might be expected in the bids. The third section discusses the contracts signed on vesting, and the generators' marginal costs. In the fourth section, the bids from coal-fired stations are compared to the estimated costs. Section 4.5 discusses the generators' open cycle gas turbines, used to meet peak demands and to support the transmission system. Their oil-fired stations, which generally run infrequently, are not discussed, in part because their costs depend on individual, confidential, oil contracts, and could not be quantified. A final section offers some conclusions.

4.2. Modeling the Pool

To understand the pool, it is almost essential to use a greatly simplified model. The pool uses well over a thousand price bids and a mass of technical data as inputs to a large linear program that minimizes the cost of meeting the demand for electricity. Even if the rules were simplified so that each set submits a single price, and the cheapest n sets then earn the nth price, mixed strategies would generally be required, and solutions for the practically important case of firms with more than one plant, similar costs, and capacity constraints, have not yet been found.

A more fruitful approach is to take the firm as the unit of analysis and assume that each firm submits a function giving a price for every level of output, from zero up to its capacity, as shown in Figure 4.1a. These functions can be added horizontally, to give an industry merit order, shown in Figure 4.1b. The intersection of the industry merit order and the demand curve for each half hour gives the level of the system marginal price. This can be inserted into the firm's function to find its output in that period. This approach ignores technical constraints and indivisibilities, but it can allow straightforward equilibria, obtained by applying Klemperer and Meyer's (1989) supply function techniques.

Their techniques have been adapted to allow for electricity contracts by Green (1992), and that paper's model of the pool is set out here. Before

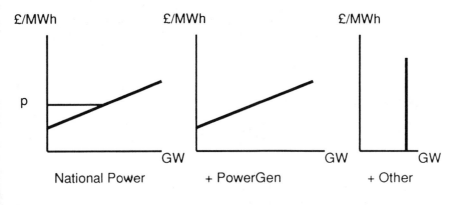

National Power   + PowerGen   + Other

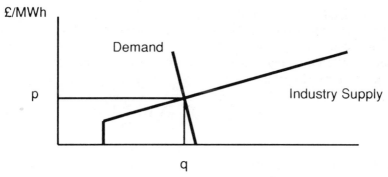

Figure 4.1. Modeling the Pool

trading in the pool commences, generator i has sold contracts for a volume x_i with a strike price of f_i. In the game, the generator submits a function $q_i(p)$, which is combined with the other generator's supply function $q_j(p)$ and the net demand[3] function $D(p, t)$ to give the market-clearing price $p^*(t)$. The generator's costs, which are assumed to be symmetric, are given by $C(q_i)$. To ensure equilibrium, we make the assumptions that $D_p < 0$, $D_{pp} \leq 0$, $D_{pt} = 0$, $C_q > 0$, and $C_{qq} \geq 0$.[4]

If the generator could sell q_i at a price of p, its profits would equal

$$\pi = p(q_i - x_i) + fx_i - C(q_i). \tag{4.1}$$

The supply function that maximizes profits is the locus of price-output combinations that maximize profits for different levels of demand. To find these points, it is necessary to replace q_i in equation (4.1) by the amount that the firm could sell in equilibrium ($q_i^*(p, t)$). That is given by the residual demand available to the two generators, less the quantity that the rival firm would supply at that price, given by its supply function $q_j(p)$:

$$q_i^* \ (p, \ t) = D(p, \ t) - q_j(p). \tag{4.2}$$

Using equation (4.2) to substitute for q_i, the first derivative of profits with respect to price is given by

$$\frac{d\pi}{dp} = (q_i - x_i) + (p - C_q(q_i)) \left(D_p - \frac{dq_j}{dp} \right). \tag{4.3}$$

The first-order condition for a maximum is that equation (4.3) should equal zero. Any pair, q_i, p, for which this holds is a point on the profit-maximizing supply function. Equation (4.2) could be used to find the value of t for which that pair would maximize profits, but this is not necessary to derive the supply function.[5] Klemperer and Meyer (1989) show that under certain assumptions we can rearrange equation (4.3) and that this will give a set of differential equations for the profit-maximizing supply function $q_i(p)$, one for each firm.

$$q_i(p) = x_i + (p - C_q[q_i(p)]) \left(\frac{dq_j}{dp} - D_p \right). \tag{4.4}$$

It should be noted that if $q_i = x_i$, so that the generator is fully contracted, profit maximization implies that price should equal marginal cost, since the second bracketed term in (4.4) is positive. If the generator has sold more electricity under contract than in the spot market, then equation (4.4) implies a price below marginal cost. On the other hand, if the generator is undercontracted, then it will wish to push the price above marginal cost. The extent to which this is profitable will depend on the expected response of demand to a change in price, and on the slope of its rival's supply function, since if these are elastic, the generator will not be able to raise the price without sacrificing a large amount in sales, whereas inelastic demand and an unresponsive rival would allow the generator to obtain a higher price with little reduction in output. The desired price-cost margin will generally rise with the level of output, since the cost of pushing up the price is balanced by the gain on intramarginal units, which rises as more are sold.

Green (1992) shows that an increase in x_i will lead to an increase in generator i's equilibrium output and to a reduction in the output of generator j and in the equilibrium price. Effectively, selling electricity under contract reserves an equivalent volume of sales through the pool for that generator. An increase in the amount of reserved sales will lead to a lower price in the smaller unreserved market, since the benefits from raising prices fall as the level of unreserved sales falls. With the rational expectations assumption that contracts can be sold for a strike price equal to the expected price in the pool, given the volume of contracts sold, the paper then shows that fierce competition in the contract market might lead generators to sell all of their output under contract, leading to low pool prices, but that if competition in the contract market is less severe, uncontracted sales, and the price-cost markup, will be larger.

Hypotheses about contracting strategy are unlikely to be testable, as contracts are highly confidential. In addition, many of the contracts signed so far were negotiated under government pressure to increase the quantity of British coal bought, which may have distorted their terms. This chapter therefore concentrates on a selection of the bids made in the pool so far, since these are in the public domain. The preceding discussion makes it seem likely that the generators would set price equal to marginal cost for the amount of electricity that they have sold under contract.[6] If the generators have too much market power, their bids will diverge from marginal cost whenever they are under- or over-contracted, and such divergences will increase with the amount by which contracted sales differ from expected output. These propositions can be tested, given some information on costs and on the contracts presently in force.

4.3. Contracts and Costs

When the electricity supply industry was privatized, the government imposed a portfolio of contracts that were designed to ensure a stable transition to the private sector. Prices to the franchise market were to rise by no more than the rate of inflation during the first three years after vesting day, and each REC's purchasing requirements for this market were covered by a set of three-year contracts with National Power and PowerGen. The quantities and strike prices in these contracts were set at the expected levels of the REC's demand for the franchise market and of the pool price for the times at which each contract would be in force. The contracts also specified lump sum payments from the RECs to the generators, to compensate them for the excess capacity inherited from the CEGB, and for the price that they were paying for their coal. The generators were required to take almost all of their coal requirements for the three years from British Coal, with take-or-pay contracts covering 70 million tons in 1990–1991 and 1991–1992 and 65 million tons in 1992–1993. The average price in these contracts, initially 180 pence per gigajoule (GJ), but indexed to the retail prices index and to the pound per dollar exchange rate,[7] was significantly above the price of imported coal, which was 120 pence per GJ in 1989–1990. In the future, the generators were likely to obtain their marginal coal supplies at world prices, and so the marginal cost of burning U.K. coal would be the future world price rather than the past U.K. price. The forecast pool prices were derived using the world price as the cost of coal and assuming that the generators would bid at marginal cost on this basis. Such bids would not cover the average cost of U.K. coal, and so part of the lump sum in the contracts between the generators and the RECs was to cover the difference.

An additional portfolio of contracts covered the nonfranchise market. The lump sums for these contracts did not include the subsidy to British Coal or for excess capacity, so that prices in this market could be lower than in the franchise market. Because customers were to be given the opportunity to change supplier after the contracts were signed, an umbrella agreement between the electricity companies allowed them to swap contracts in order to keep their portfolios and their expected supplies in balance. These nonfranchise contracts were due to expire on March 31, 1991, after one year. Finally, the generators signed one-year contracts to cover the expected demands of some of the largest industrial consumers at prices that ensured that those consumers did not face a real increase in their electricity price. Those prices were significantly lower than the prices in any other contracts, and they were backed by contracts with Nuclear Electric at the same prices, so that the state-owned firm covered the cost of this subsidy.

Figures for the amount of electricity sold under contract were given in the generators' prospectus (Kleinwort Benson, 1991). In sum, National Power, with a total declared net capacity of 29.5 GW in January 1991 (of which around 90 percent would be available on a typical winter day), held a portfolio of contracts and direct sales with a maximum net capacity[8] of 25.8 GW. PowerGen, with a declared capacity of 18.8 GW, had contracts and expected direct sales with a maximum net capacity of 16.6 GW. Many of the contracts were due to expire at the end of March 1991, but National Power would still be liable for 21.5 GW of contracts with the RECs, and PowerGen for 13.1 GW. All of their direct sales were for a single year in the first instance, and so were due to expire during 1991, but many of them will have been renewed, and the generators may have acquired some new customers. It is therefore difficult to know the level of the generators' peak contract and direct sales liabilities after the winter of 1990–1991. The prospectus gives no information on the level of contract sales at times away from the peak, although it seems likely that the relationship between the vesting contracts and the forecast level of demand would be similar across most time periods, implying a high level of contracting in 1990–1991, falling slightly for 1991–1992 and 1992–1993.

If the generators were almost completely contracted during the first year after vesting, we would expect them to set their bids close to the level of marginal cost. If they became undercontracted in the following year, failing to renew some of the contracts that expired in March 1991, then they might wish to bid above marginal cost, particularly with peaking plant. To assess whether their bids followed this pattern, we need to assess the level of the generators' marginal costs.

In principle, marginal costs can be computed quite easily as a station's cost of fuel per kWh (of energy), divided by the station's thermal efficiency (expressed as a fraction). One potential problem is that thermal efficiency will depend on a station's operating regime, for a station that is often turned on and off will incur the significant fixed costs of heating up the station more often, and hence have a lower thermal efficiency, than a station that is run continuously. The solution used here was to treat all stations on a common basis, ignoring these differences. The CEGB's annual reports gave information on annual thermal efficiencies, and the figure adopted for each station was one from early in its life, when it was operating with a high load factor at a high thermal efficiency. The price vectors that each generating set submits were converted into bids on a consistent basis, assuming that each set would be turned on and then operate for sixteen hours at full capacity,[9] implying a (relatively high) load factor of 66 percent. The unweighted average of the bids from each of the station's sets was then used to represent the station's bid. The pricing mechanism ensures that a station's bid is equal to the average cost per kWh produced during a period of operation, and so it is appropriate to compare these bids with the historic average thermal efficiencies.

The second element in the marginal cost calculations, the cost of fuel, is harder to pin down. The bulk of the generators' coal supplies during 1990–1993 will have been obtained under their contract with British Coal, at an average base pithead price which started at 180 pence per GJ, and rose to 187 pence per GJ by 1992–1993. In 1990–1991, and in the first half of 1992–1993, the exchange-rate provisions in the contract will have been triggered with an average exchange rate of around $1.90 per £1, reducing the coal price to some extent. It would be possible to ignore these adjustments and to add transport costs to the base price to give a result that would be close to the average delivered cost of coal, given the low volume of imports at present. However, because the British Coal contract was on a take-or-pay basis, with a minimum take that, with existing stocks, was well above the generators' needs, coal supplied under the contract has never been marginal.

The generators would replace their marginal coal burn with imported supplies (if the station was within reach of a port), or with supplies from British Coal bought after the initial contracts expired.[10] In 1989–1990, the CEGB paid a delivered price of 120 pence per GJ for imported coal, although this only applied at the few stations that were most favorably located to take imports. The world price of coal can be observed, and has not changed greatly since 1990, so that 120 pence per GJ could be regarded as a lower bound for the marginal cost of coal to the generators. On the other hand, the total import capacity available to the generators is significantly lower than their demand, and many of their inland stations would incur a substantial transport cost premium were they to take their coal from abroad. For those stations, the relevant marginal cost would be the price per GJ that was expected to apply to marginal supplies from British Coal after March 1993. During 1992, the generators were negotiating with British Coal over the terms of a five-year coal contract, and with the RECs over corresponding electricity contracts, attempting to set up a back-to-back deal that would guarantee significant parts of their costs and revenues. Leaks from those negotiations implied that the pithead price in the first year of the contract would be approximately 150 pence per GJ, declining to around 130 pence per GJ by the last year. Thus, it would be reasonable to take a price of 150 pence per GJ as an upper bound on the cost of marginal coal that was expected in September 1992, when the sample ends, and earlier expectations were probably similar. Subsequently, British Coal's announcement that it would have to close thirty one of its fifty remaining pits because the new contracts involved much lower tonnages (finishing at 25 million tons a year) sparked off a political storm. The government was forced to announce a review of the decision, which did not affect the final outcome, but the uncertainty over its results could not have affected the generators when nobody expected the review.

If we add transport and handling costs[11] of around 10 pence per GJ on average to this price, we would have an average marginal cost for British coal of about 160 pence per GJ. Imported coal would be significantly cheaper at the docks, but only a few stations could take it without incurring additional transport costs, and if we take the same figure of 10 pence per GJ for these, we could assume a price of 130 pence per GJ for imported coal delivered to the more accessible inland stations.

In the graphs which follow, the lower line represents marginal costs with a fuel price of 130 pence per GJ, and the upper line marginal costs with a fuel price of 160 pence per GJ. Some stations would be able to

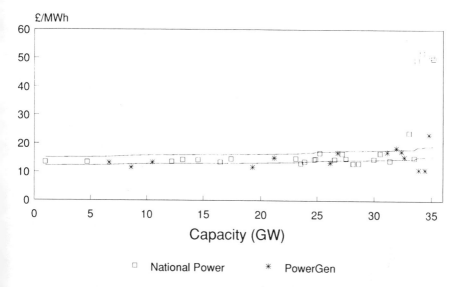

Figure 4.2.   Coal-Fired Merit Order (October 3, 1990)

obtain marginal fuel supplies at lower prices, if they were particularly close to mines or to import facilities; conversely, others would face higher delivered prices because of higher transport costs. These stations might be identified, but it would be difficult to quantify the effect on their costs from information in the public domain. In this chapter, no attempt is made to quantify these effects, and so the marginal cost curves should only be taken as approximations that will not be completely accurate for individual stations.

4.4. The Observed Bid Prices and Estimated Costs

Figures 4.2 through 4.6 show the bids and the estimated marginal costs for the coal-fired stations owned by National Power and PowerGen on five Wednesdays[12] between October 1990 and September 1992. Similar patterns to those reported here were obtained for other days studied. The vertical scale is the same in all figures, to ease comparison, but the horizontal scale differs, reflecting changes in the capacity available. Because it is impossible to obtain accurate station by station costs, a bid that is only just outside the range of marginal costs shown here should not be taken as

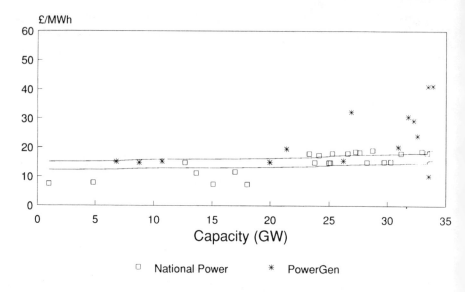

Figure 4.3.   Coal-Fired Merit Order (February 13, 1991)

proof that the station was bidding above or below the level of its costs, but a large difference would imply that it probably was.

The prices for the first day in the sample, October 3, 1990, are almost all close to the band of estimated marginal cost. The two companies were still wholly owned by the government, and it is unlikely that they would have wished to disturb the market in the immediate run-up to the RECs' flotation the following month. Given these factors, and the large volume of electricity sold under contract, bidding at marginal cost seems a plausible strategy. To some extent, the fact that almost all of the stations had bids in the region between the two marginal cost lines is a useful test of the level of marginal cost, confirming that the estimates are plausible. A few small National Power stations have submitted higher bids, which would force up the price if demand was very high, possibly above the company's contracted sales. Two small PowerGen stations submitted very low bids. These stations were closed a few months later, and may have been bidding low to ensure that all of their coal stocks were burnt before they closed.

Figure 4.3, for February 13, 1991, shows bids made near the winter peak.[13] National Power now seems to be bidding its larger stations at below marginal cost, although it is quite possible that the level of demand was so high that these stations would always be intramarginal, and would never set the price. Some of PowerGen's smaller stations are setting prices

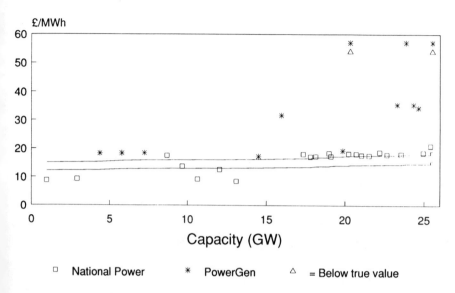

Figure 4.4.   Coal-Fired Merit Order (June 26, 1991)

some way above their estimated costs. The National Power stations that bid high in October 1990 were closed, while the other small stations mostly bid at around the level of their estimated marginal costs. This would be consistent with the profit-maximizing strategies outlined in Section 4.2 if National Power was overcontracted and PowerGen was undercontracted, so that the former wanted to push the pool price down and the latter to raise it. We do not know the generators' positions, but since the figures in Section 4.3 implied that, relative to its capacity, PowerGen had slightly fewer contracts than National Power, we should be wary of accepting the theory.

Almost all of the National Power stations, and the larger PowerGen stations, submitted similar bids on June 26, 1991, shown in Figure 4.4. There is a noticeable change, in that more of PowerGen's smaller stations are now bidding significantly above the level of their marginal costs. PowerGen was known to be significantly undercontracted at this time, while National Power was more fully contracted, so that the theoretical explanation of their bidding strategies fits the facts.

Two PowerGen stations submitted bids that were off the edge of the graph. Ferrybridge B (at 20 GW) submitted a bid of £101 per MWh, and Hams Hall (at 25 GW) bid £109 per MWh. The explanation for these bids can be found in the regulator's *Report on Constrained-on Plant* (OFFER,

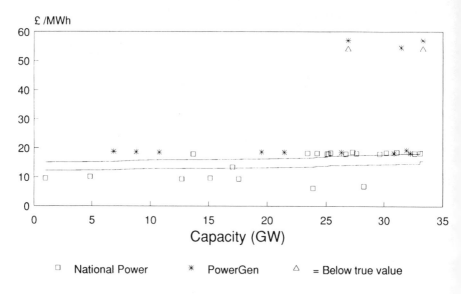

£ /MWh

Figure 4.5.   Coal-Fired Merit Order (February 12, 1992)

1992b). Each station was directly connected to a (separate) regional distribution system that was only connected to the national grid at a single substation. Neither substation had enough spare capacity to guarantee that supplies could be maintained in the event that two of the transformers at the substation failed simultaneously, and so the power stations had to operate to provide the required level of security. When stations are constrained in this manner, they are paid their own bid price, rather than the pool price, and so PowerGen had an incentive to raise the stations' bids. A limit pricing argument might imply that the station should not abuse this local monopoly so much that the National Grid Company feels obliged to build additional transformers and remove the constraint, but because PowerGen planned to close both stations in any case, additional transformers were already under construction. Furthermore, the costs of constrained on running were then borne by all electricity suppliers (and hence their customers) and so NGC had no financial incentive to remove constraints.[14] It was given such an incentive in 1994.

The same basic pattern is repeated in Figure 4.5, showing the bids made on February 12, 1992. A few PowerGen stations that were bidding in the range from £30 to £40 per MWh have reduced their bids to approximately the level of their estimated marginal cost. Mild winter weather, and increasing output from Nuclear Electric, may have reduced the market available

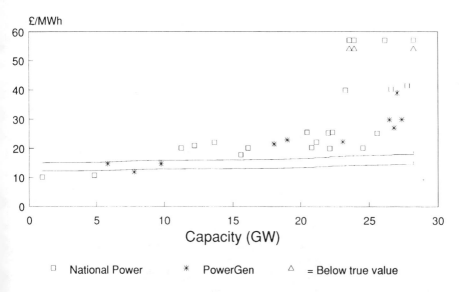

Figure 4.6. Coal-Fired Merit Order (September 16, 1992)

to the company, so that its effective contract cover was higher, giving it an incentive to bid close to marginal cost. Another new feature, albeit one observed on other days not reported here, is that two of National Power's midmerit stations are bidding low prices. Blyth A, at 24 GW, bid £6.14 per MWh, and Blyth B, at 28 GW, bid £6.74 per MWh. One possible aim of this strategy may have been to reduce coal stocks near the end of a contract year. An alternative possibility, since these stations are relatively isolated in the northeast, is that they might have faced transmission constraints under which they were unable to export their power. The pool rules specify that they would then be paid the difference between the system marginal price and their own bid as a lost profit payment, giving them an incentive to bid low and maximize this payment.

Figure 4.6 shows the bids for the final day considered here, September 16, 1992. The pattern has changed, to some extent, for while a few large stations are still bidding at marginal cost, almost all of the other stations have submitted higher bids. In general, the stations with the highest cost have the highest price-cost margin, which is consistent with the theory. It is at least possible that because the generators were negotiating their contracts for the next five years, they were attempting to push up the pool price to support a higher future contract price, despite the restraining effect of their existing contracts.

The two PowerGen stations that submitted very high bids closed at the end of March 1992, and a group of National Power stations have set high bids. Again, these stations are discussed in the *Report on Constrained-on Plant*. The first three. Aberthaw A (£65 per MWh), Uskmouth (£67 per MWh), and Rugeley A (£60 per MWh) are in areas (South Wales and the West Midlands) with more demand than generation, and transmission constraints mean that some stations in those areas are required to run. National Power had tended to submit low bids from these stations when demand was high, so that they could generate normally, earning SMP, but when demand was lower, the stations would be too far down the merit order to generate normally. The transmission constraints still applied, and because National Power was able to set high prices for most of the stations that could meet them, those stations were able to earn significant sums for constrained running. The final station, Skelton Grange, bid the rather higher price of £97 per MWh. In the summer of 1991, work on a nearby transmission circuit meant that the station had to run, and it bid a high price during the few weeks involved (OFFER, 1992b, pp. 25–26). It is probable that other transmission work in the area, which took place at this time, gave the station another temporary local monopoly, and that it took advantage of it. The temporary nature of the constraint means that to earn the same annual revenue as the other stations, Skelton Grange would have to bid a higher price. National Power has declared a policy of setting bids to earn revenues equal to each constrained station's cost and contribution to overheads (OFFER, 1992b, pp. 42–57), and the pattern of these bids seems consistent with such a policy.

4.5. Open-Cycle Gas Turbines

National Power and PowerGen also own 3.5 GW of open-cycle gas turbines, derived from aero engines. Some are auxiliary turbines, located at large coal- or oil-fired stations, which are used to meet peak demands and would allow those stations to restart generating independently in the event of large-scale grid failures. Other, larger, turbines are in purpose-built stations, mainly in the south, and are used to relieve transmission constraints as well as meeting peak loads. The turbines, which have high running costs, practically never enter the main generation schedules[15] and are compensated on the basis of their own bids, rather than by setting SMP. The stations tend to submit high bids, although, when considered alongside their high costs and their infrequent operation, they may not be making excessive profits.

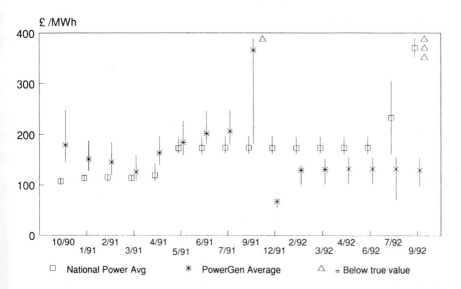

Figure 4.7. Gas Turbine Bid Prices (Selected Days, 1990–1992)

Figure 4.7 shows the average (capacity-weighted) bid submitted by each company's stations on each of sixteen days between October 1990 and September 1992, together with the highest and lowest bids. National Power's stations have almost always submitted similar prices, while those of PowerGen have tended to be more dispersed. At first, National Power's prices were lower, although PowerGen's declined gradually to meet them. For much of April 1991, PowerGen declared its turbines to be unavailable, arguing that they were unable to cover their costs under the present pool rules. At the end of the month (the data point is for April 30), PowerGen returned its plant to service at a higher price, and National Power had also raised its prices by May 22, the next data point. For more than a year the observed prices for National Power stayed constant, while PowerGen started by bidding higher prices, and then bid lower, at a fairly constant level from February 1992. PowerGen's average price for September 9, 1991 was raised by three small auxiliary turbines that bid £857 per MWh, while the bids for the company's other stations were similar to the previous set examined. The last two sets of bids submitted by National Power were significantly higher than the company's earlier bids: the last bids, for September 16, 1992, were all in excess of £1000 per MWh. It is possible that the publication of a *Report on Gas Turbine Plant* in June (OFFER, 1992a), which agreed

24 July 1991

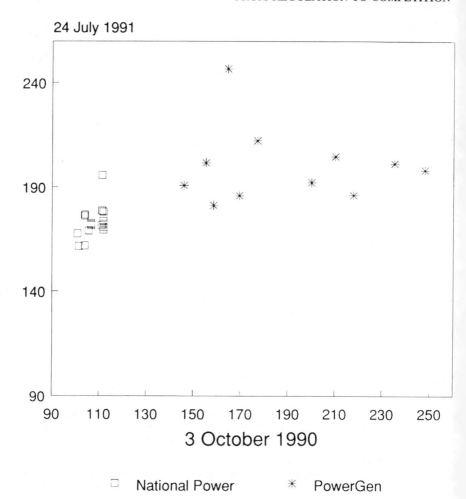

3 October 1990

▢   National Power   ＊   PowerGen

Figure 4.8.   Gas Turbine Bid Prices (Pounds per MWh)

that some gas turbines were not covering their costs, may have been seen as a green light for higher prices.

It is unlikely that the changing bidding strategies for these stations have been driven by changes in their costs, which would generally have affected both companies in the same way. Figure 4.8 presents individual stations' bids for two days as a further test to see how likely it is that changes in costs have caused the changes in bids. It is likely that all the stations' costs would change by similar proportions if fuel prices changed, so that their relative prices should not change, and the points on the scattergraph would form a diagonal line sloping up to the right. Since there is little correlation between the bids made on October 3, 1990 and on July 24, 1991 for National Power, and practically none for PowerGen, it is extremely unlikely that both days' bids were cost reflective.

4.6. Conclusions

This chapter has examined the behavior of the two major generators since the restructuring of the electricity supply industry in England and Wales. The government established an unregulated market for wholesale supplies of electricity, trusting to competition to keep prices in line with costs. The experience to date can only be regarded as a partial test of whether this trust is well founded, for the generators' incentives have been affected by contracts signed before privatization, which were designed to ensure stable electricity prices in real terms for three years. From April 1993, once these contracts have been renewed by private firms, and the new contracts start to affect behavior in the pool, we will be able to draw more conclusions.

In the meantime, most of the larger generating sets have bid prices that are tolerably close to the estimated level of their marginal costs. Some sets have bid lower prices, but if they ran continuously, they may not have affected prices at the margin. Many of the smaller sets have bid prices that are above their estimated marginal costs, and these bids will tend to have raised the system marginal price. A few stations have exploited local monopolies to set prices far above their marginal costs, which moves them too far down the merit order to operate when the constraint does not apply. When the constraint binds, neither system operation nor the system marginal price is distorted, but the cost of electricity to the consumer is increased by the additional uplift payments. In the short term, the price of electricity will be too high, while if plants that appear at the margin are in the wrong place in the merit order, the cost of generation will be raised. In the longer term, this gaming does not inspire confidence that the industry will be sufficiently competitive to ensure that bids are set equal to costs and provide an efficient outcome.

This is not to say that electricity prices will rise far above costs as soon as the present contracts expire. The industry has a regulator with substantial powers, and although those powers are ill defined with respect to the generators, fear that they will be exercised is likely to persuade the duopolists to keep prices down to an acceptable level. National Power's policy of setting prices for constrained-on plants that cover their costs, rather than maximizing profits, is an example of this. It is perhaps regrettable that these low prices will be the result of companies having to guess how much they could get away with, rather than the outcome of clear regulatory rules, or of a competitive system that did force prices down to costs.

Appendix: The Pool

A special trading system had to be designed to allow competing generators to connect to the same electrical network. Electricity cannot be stored, and if the inflows to the system do not balance the outflows almost exactly, widespread failures can be expected. This means that generators must follow the grid controllers' instructions exactly and respond to unexpected changes immediately. There would be no time to agree a price for such reactions, and so rules must be established in advance that specify the payments to be made in all eventualities. If the generators know that they will be paid appropriately, then they will be happy to submit to the dispatchers' operational control. The electricity pool achieves this by asking each generator to submit prices each day and then ensures that each station will receive at least its own price whenever it operates. In normal operation, all sets are paid the system marginal price (SMP), which is the highest price of any set in normal use, while other, more expensive sets that are operating to meet constraints are paid their own price.

The second requirement that the pool system had to meet was to continue the system of merit order dispatch, which minimizes the total cost of generating by choosing the marginal station in use so that its running costs are lower than any station that is not in use but capable of meeting demand. The pool does this by ranking the stations in price order, using the information in their bids, and then selecting those stations with the lowest prices, subject to technical constraints on their operation. One important constraint is that large generating sets cannot be turned off for short periods and that, even if it is possible to do so, it may be cheaper to continue generating on part load than to turn off and incur the cost of heating the set from cold a few hours later. At times when this constraint applies, known as Table B periods, the marginal cost of electricity, and hence SMP, is given by the marginal station's bid for producing an additional unit.[16] At other times, changes in the demand for electricity would be met by turning sets on or off, and so all the costs of running the marginal set should be considered to be marginal. The price in these Table A periods therefore includes the cost of starting up the set and of running it with no load, as well as the cost of each unit generated. This total cost is divided by the number of units that it is scheduled to generate before it is next turned off to give an average cost per unit generated, used as SMP. A large computer program, known as GOAL, uses these rules and the generators' bids to produce an operating schedule for the following day and forty-eight values of SMP, one for each half hour.

A second element in the pool price rewards generators for providing capacity, whether or not they are scheduled to generate, since the system must carry enough capacity to meet the risk of high demand coinciding with plant failures, and some of that capacity would rarely generate. The capacity element is defined as the loss of load probability multiplied by the value of lost load less the greater of SMP and the set's own bid price. That gives the net value of the energy that the set could provide, multiplied by the probability that the set would be required to provide it. The loss of load probability is calculated from data for demand and available capacity, subject to some averaging to reduce volatility, while the value of lost load is governed by the regulator. It was set at £2 per kWh in 1990 and indexed to the retail price index but may be increased, or decreased after giving notice, if the incentive to provide capacity needs to be strengthened or weakened.

The sum of SMP and the capacity payment for that half hour give the pool input price, paid to generators for the output that they have been asked to produce in GOAL's unconstrained day ahead schedule. Because the prices, and the schedule, are produced a day in advance, this part of the pool could be seen as a forward market,[17] in which the majority of actual transactions are covered the day before they take place, leaving only a small margin for spot transactions. Spot transactions are required because some stations that are scheduled to generate may break down, so that others must take their place, and the level of demand may differ from that forecast when the schedule was compiled. Furthermore, GOAL produces an unconstrained schedule, which the transmission system may not be able to accommodate, so that some sets may be constrained off (in exporting areas) and others constrained on (in importing areas) to reduce the flows over the system. This means that many stations will generate more or less than in the day ahead schedule used to calculate SMP. Stations that generate less than in the schedule must buy back their excess forward sales, while those which generate more can sell the uncovered electricity. These spot transactions are generally made at the station's own bid price, rather than SMP, with deductions made for stations that did not follow the controllers' orders. The net cost of the transactions, together with capacity payments made to stations that were not generating at the time, and payments made by NGC for various ancillary services required to keep the transmission system in electrical balance, are spread over all the units generated in Table A (nontrough) periods, to give an amount per kWh known as uplift.

The pool output price, paid by all electricity purchasers, is therefore equal to the pool input price (SMP + capacity) in trough periods and to the sum of the pool input price and uplift in Table A periods. Metered demands in each half hour are scaled up by the average transmission losses so that they equal metered generation, and the sums paid into and out of the pool for every day will balance exactly.

This paper was written in 1993. Pool prices increased significantly in the years following privatisation, causing growing concern, and the regulator's reports on the issue acknowledged that 'National Power and PowerGen together had market power and exercised it in a significant way'. In 1992, the regulator had committed himself to deciding by 1995 whether the companies should be referred to the Monopolies and Mergers Commission (MMC), which would conduct an investigation and order remedies for anti-competitive behaviour. In July 1993, he responded to the most recent increases in the pool price by announcing that he would study the generators' costs and profits, to see if these were excessive, and would justify an MMC inquiry. He aimed to make his decision by the end of the year.

In December, he announced that a deal to avoid an inquiry might be possible, and in February 1994, its terms were announced. National Power agreed that it would try to sell 4 GW of plant, and PowerGen 2 GW. These sales were intended to increase competition among 'mid merit' power stations—the baseload market now has enough competition from nuclear stations and new entrants. To help consumers while the sales were negotiated, the companies promised to bid in such a way that the average pool price would not exceed £24 per MWh (in October 1993 prices) for the next two years. This is based on the regulator's view of the prices that a competitive market would produce. The companies now know what they can get away with.

Notes

1. This research was supported by the Economic and Social Research Council grant R000231811, *Privatisation and Re-regulation of Network Utilities*. I have gained from many discussions with David Newbery, and from and Nils-Henrik von der Fehr and David Harbord's comments (1993) about an earlier paper (Green, 1991) on this subject.
2. There is no requirement for the parties to a contract to be electricity companies (or

large consumers who buy in the pool), but practically no contracts have involved outsiders to date.

3. This demand function is net of supply by Nuclear Electric and independent generators, who are all assumed to bid at marginal cost, well below the duopolists' prices.

4. The assumptions about C are natural, since the generators can choose which sets to run (via their bids) and can ensure that the cheaper sets are higher in the merit order. The assumptions on the second derivatives of D are less well founded, but they are sufficient rather than necessary conditions for equilibrium.

5. That is, as long as $D_{pt} = 0$. If the slope of the demand curve depends on the time of day, then equation (4.3) is no longer independent of time, and this must be taken into account when deriving the profit-maximizing supply function.

6. An alternative hypothesis, which would imply greater market power, is that the generators would set prices above marginal cost in any case, forgoing present profits to push up expectations of future pool prices, raising the price at which they can sell contracts and hence their future profits.

7. The coal price was to rise by 5.5 percent less than the increase in the RPI in April 1991, and by 5 percent less in April 1992. If the pound per dollar exchange rate was outside the range from $1.40 to $1.75 per £1, the coal price would be adjusted accordingly.

8. That is, National Power had sold contracts for differences and expected direct sales of 27.6 GW but also held contracts for differences (from PowerGen and Nuclear Electric, swapped when some large industrial consumers started to buy from National Power) for 1.8 GW.

9. This bid, the Table A genset stack price, was equal to the start-up charge plus sixteen times the no-load price, divided by sixteen times the set's capacity, plus the set's incremental kWh price (or their weighted average, where the set had more than one).

10. National Power has stated that it uses the international price of fuels when calculating station costs as part of the process of setting its bids (OFFER, 1992a, pp. 53–54).

11. The generators' prospectus also states that the average delivered price of British Coal was 197 pence per GJ in 1989–1990. The pithead price was about 5 percent higher in 1989–1990 than in 1990–1991, which would imply a level of around 190 pence per GJ in 1989/90 and average transport costs of just under 10 pence per GJ.

12. There has been some evidence (OFFER, 1992b, pp. 21–41) that the bids submitted on weekdays are different from those made at weekends, when demand is lower, relative to the amount of capacity needed for the weekly peak. This sample only uses bids from Wednesdays to ensure that any effects of this sort do not distort the results.

13. Electricity demand in the United Kingdom peaks in winter evenings, when heating, lighting, and cooking demands occur before industrial and commercial loads have started to fall. Summer loads are much lower; in particular, there is little need for the air conditioning that gives the peak in the southern States.

14. It should be noted that NGC has installed additional equipment to reduce a constraint on Fawley station, in the South of England, even though NGC's direct gains from this expenditure were small.

15. In September 1991, a quirk in the scheduling programmes meant that a gas turbine was included in the main schedule for an hour and a half, sending the price up to £160 per MWh. Those large consumers who were able to reduced their demand so much that the station did not need to run, but the price had already been set, and the episode was the final straw leading up to the regulator's inquiry into pool prices (OFFER, 1991). The (over-contracted) owner of the station lost money from the high price.

16. They are allowed up to three unit prices, for different levels of output, so that some sets bid a low price (or zero) for much of their capacity, and higher prices for the second and

third tranches. This usually ensures that the set will generate at least to the edge of the low-price tranche, which avoids the problems of running with very low loads or turning off, while higher prices will set SMP.

17. This view is expounded by Hunt and Shuttleworth (1993).

References

Bolle, F. 1992. "Supply Function Equilibria and the Danger of Tacit Collusion: The Case of Spot Markets for Electricity." *Energy Economics* (April): 94–102.

Department of Energy. 1988. *Privatising Electricity*, Cm 322, London: HMSO.

Green, R.J. 1991. "Bidding in the Pool." Mimeo, University of Cambridge. Paper presented at the OFFER Seminar on Pool Prices, October.

Green, R.J. 1992. "Contracts and the Pool: Competition in the British Electricity Market." Mimeo, University of Cambridge. Paper presented at the Seventh Annual Congress of the European Economic Association, Dublin, August.

Green, R.J., and D.M. Newbery. 1992. "Competition in the British Electricity Spot Market." *Journal of Political Economy* 100(5), (October), 929–953.

Helm, D., and A. Powell. 1992. "Pool Prices, Contracts and Regulation in the British Electricity Supply Industry," *Fiscal Studies*, 13(1) (February), 89–105.

Hunt, S., and G. Shuttleworth. 1993. "Forward, Option and Spot Markets in the UK Power Pool," *Utilities Policy* 3(1), (January), 2–8.

Kleinwort Benson Limited. 1991. "Main Prospectus: National Power plc & Power-Gen plc Offers for sale by Kleinwort Benson Limited on behalf of the Secretary of State for Energy." London: Kleinwort Benson Limited.

Klemperer, P.D., and M.A. Meyer. 1989. "Supply Function Equilibria in Oligopoly Under Uncertainty." *Econometrica*, 57(6) (November), 1243–1277.

OFFER. 1991. *Report on the Pool Price Inquiry*. Birmingham: Office of Electricity Regulation.

OFFER. 1992. *Review of Pool Prices*. Birmingham: Office of Electricity Regulation.

OFFER. 1992a. *Report on Gas Turbine Plant*. Birmingham: Office of Electricity Regulation.

OFFER. 1992b. *Report on Constrained-on Plant*. Birmingham: Office of Electricity Regulation.

Powell, A. 1993. "Trading Forward in an Imperfect Market: The Case of Electricity in Britain." *Economic Journal* 103(417) (March), 444–453.

von der Fehr, N-H M., and D. Harbord. 1992a. "Long-Term Contracts and Imperfectly Competitive Spot Markets: A Study of the U.K. Electricity Industry." Memorandum No. 14 of 1992, Department of Economics, University of Oslo.

von der Fehr, N-H M., and D. Harbord. 1993. "Spot Market Competition in the U.K. Electricity Industry." *Economic Journal* 103(418) (May), 531–546.

5 TRANSMISSION PRICING IN ENGLAND AND WALES

Sally Hunt and Graham Shuttleworth

When the old Central Electricity Generating Board (CEGB) for England and Wales was broken up in 1990 into its constituent parts, the high-voltage transmission company—the National Grid Company (NGC)—had to develop a set of transmission charges for the generator companies and their customers. In the excitement over the creation of a competitive market in electricity, the basis for transmission charges received relatively little attention. As a result, problems arose within a year or two of privatization,[1] mainly from the inadequate differentiation of charges by location and from the treatment of transmission constraints.

NGC is now trying to rectify the deficiencies of the original system but is predictably encountering resistance to the proposed changes. In areas where charges are increasing, of course, existing users lose out and complain vociferously. The plaudits of those whose charges fall seem to count for less, in any regulatory debate.

NGC's only alternative appears to be to slow down the speed of tariff changes, in the hope that tariffs eventually reach the desired levels without disturbing existing users—but also without, in the mean time, encouraging new users to locate their business in the wrong place. A better solution might be to allow NGC to charge different prices for existing and new

97

users. Instead of viewing this as price discrimination, it should be regarded as the outcome of the normal process of negotiating long-term contracts for transmission, whose price changes over time. As we discuss below, selling transmission on long-term contracts will have other benefits and deserves further investigation.

5.1. The Current Structure of Electricity Transmission Charges

NGC operates the National Grid, a network of high-voltage lines (275 kV and 400 kV) that covers the whole of England and Wales. The National Grid links generators with the low-voltage networks of the regional electricity companies (RECs), which engage mainly in local distribution and retail supply of electricity. Some generators are connected to the low-voltage networks, but they affect net flows on the National Grid just as if they were connected directly to a high-voltage line. NGC takes this into account when levying charges.

NGC's charges fall under two basic categories:

- *Use-of-system (UoS) charges* pay for the right of a generator or customer to transmit electricity across the high-voltage network; the total volume and structure of these charges are regulated by the Office of Electricity Regulation (OFFER).
- *Connection charges* pay for the assets required to connect any generator or customer to the high-voltage network; these charges are not regulated.

NGC maintains a strict delineation between generation and customers in setting and levying both of these charges.

Use-of-system charges differ according to whether the user is a source of generation or load; charges are applied to gross generation and gross load, and one may not be offset against the other to reduce charges.[2] Generators pay about 25 percent of use-of-system charges, and the other 75 percent is borne by customers.

Connection charges are divided between *entry charges* (for generation) and *exit charges* (for load); charges relate to actual assets installed, but NGC tries to maintain a similar distinction between assets for entry and assets for exit. For example, generators pay an entry charge and also an exit charge for the assets required to feed a power station's demand for electricity at any time when it is not generating enough to meet its own load from in-house sources.

The charges for generators and customers are made up as follows:

- Charges for generators

  The *entry charge* is equal to the annualized cost of connection assets for generation (pounds per year).

  The *infrastructure charge for use of system* is defined in terms of pounds per kW of connected generation capacity and pence per kWh for each unit of electricity actually sent out. This charge varies from region to region, with lower charges in the south, where there is excess demand, than the north, an area of surplus generation. There are eleven regions or zones. In 1990–1991, the range of zonal variation was from £0.00 per kW and 0 pence per kWh, to £3.13 per kW and 0.0249 pence per kWh.

- Charges for customers

  The *exit charge* is equal to the annualized cost of connection assets for customers (pounds per year). (This is also charged for power station load).

  The *infrastructure charge for use of system* is defined in terms of pounds per kW for the customer's average demand in the three half hours of actual system peak (the "triad"). This charge varies from region to region, with lower charges in the north, an area of surplus generation, than in the south, where there is excess demand. The same eleven zones apply as for generation charges. In 1990–1991, the zonal charge varied from £5.95 per kW to £8.00 per kW.

  The *system service charge for use of system* was levied as a uniform pounds per kW supplement to the infrastructure charge. The charge was £3.372 per kW in 1990. Since the services paid for by this charge were not fully specified, the reason for levying this charge was never entirely clear. It is to be abolished during 1993.

NGC's total use-of-system revenues have been regulated by OFFER since 1990. The current basis is an annual limit on total revenues, derived as a product of two variables: a price (pounds per kW) indexed to the retail price index (RPI); and the level of each year's peak demand in kW. This rule applies from April 1990 to March 1993.

From April 1993, the allowable revenue total will be equal to a price in pounds per kW, indexed to the RPI less 3 percent per annum, and multiplied by a forecast of peak demand set out in NGC's Seven Year Statement

Table 5.1.   Summary of Transmission Charges (in millions of pounds, for 1990–1991)

	Connection	Use of system: infrastructure		Use of system: system service	Total
	Entry and exit	Capacity	Energy		
Generators	£76	£97	£32		£205
Customers	£200	£325[a]		£161[b]	£686[c]
Total	£276	£422	£32	£161	£891

a. Includes £7 million paid by generators for power station load.
b. Includes £2 million paid by generators for power station load.
c. Includes £9 million paid by generators for power station load.

for 1991. These two mechanisms are peculiar to the U.K. form of regulation. RPI-X is the basic form of price regulation, applied to all regulated industries, which indexes prices to inflation (RPI) minus a factor, X (in this case 3 percent), to reflect required productivity gains. The use of a demand forecast removes any incentive for NGC to encourage the growth of system peak demand in order to increase revenues: it is similar to the ERAM mechanism adopted in some states in the United States. The annual size of the charges themselves is summarized in Table 5.1 for the year 1990–1991.

5.2. Transmission Tariff Proposals for 1993–1997

In June 1992, NGC issued proposals for the development of transmission tariffs over the next review period, then defined as 1993–1996. (National Grid Company, 1992). Much of the consultation document concerns the proposed method of calculating the tariffs, which is described as investment cost-related prices (ICRP). The key outcomes of these proposals are

- The number of zones is to be increased from eleven to fourteen, to allow transmission tariffs to reflect costs more accurately;
- The range of use-of-system charges for customers (infrastructure and system service combined) is to be widened; in 1992–1993, they varied from £9.70 to £14.1 per kW of demand at system peak; by 1995–1996, the range is to be from £0.00 to £22.1 per kW;
- The range of use-of-system charges for generation (kW and kWh charges combined, assuming a 50 percent load factor) is to be widened;

in 1992–1993, the range was from £0 to £4.7 per kW of registered generator capacity; by 1995–1996, generator's charges would vary from *minus* £7.9 to *plus* £8.5 per kW.

OFFER subsequently modified these proposals (OFFER, 1992b), in order to blunt the impact of the changes:

* The review period was extended, so that the final charges are not introduced until 1997;
* The maximum allowed increase in charges to customers was £6 per kW; this effectively halved the increase in charges to customers in Devon and Cornwall, in the southwestern peninsula of England;
* The maximum allowed increase in generator charges was £3 per kW; this affected some generators in the far north of England, and Scottish companies selling energy into England via the electricity pool;
* On-site load was to be netted off against generation, where a customer operated a small generator on the same site; this was NGC's first departure from the principle of treating generation and load separately, whenever possible.

The combined result of these changes was a gradual sharpening of the differentials between zones, although not by as much as NGC had proposed, with the effect of increasing the incentive to locate generation in the south of the country and load in the north. This represented a concerted effort to change the incentives faced by grid users to invest efficiently when locating both generation and load.

5.3. Treatment of Transmission Constraints

The differential in charges is justified with reference to the costs of providing transmission to move electrical energy around the country in a secure and stable manner. Obviously, the construction of transmission lines is the major component of NGC's costs and NGC uses the ICRP concept of incremental investment to allocate charges to different zones. However, the electricity pool's experience of operating an electricity market has demonstrated that transmission incurs further costs, which are not reflected in the costs of the lines alone.

Transmission systems are rarely designed to handle all the energy flows required by a strict, least-cost operation of a fleet of generators. It is often cheaper to reduce the net flow over some parts of the transmission system by running generators at the receiving end of the line, even if they are more expensive than other possible sources of energy. Such generators are termed *out-of-merit* (because they are not dispatched in accordance with the merit order of marginal costs), or *constrained on* (because they are dispatched to satisfy a constraint on the transmission system).

Whenever such generation is available at the receiving end of a transmission link, the transmission company may decide to build a smaller link, at lower cost. The smaller link may be cheaper in the long run, even after allowing for the costs of running out-of-merit generators whose marginal costs exceed the market value (*system marginal price*) of any energy that they produce. These excess costs should really be viewed as a cost of operating or supporting the transmission system.

Under the pooling system for England and Wales, however, these excess costs are not treated as a cost of transmission but are charged to customers instead as a uniform levy on sales, referred to as the *uplift*, without regard to their use of the transmission system. The scale and allocation of these costs has provoked many complaints, and in October 1992 OFFER produced a "Report on Constrained-On Plant" that analyzed the source of such costs and proposed alternative solutions. One proposal is to view the "out-of-merit costs of constrained-on generators" as a cost of providing secure transmission, to be charged to NGC and recovered via use-of-system charges.

NGC has indicated a willingness to take over some costs from customers of the electricity pool ("Power Plans Aim for Stable Prices," 1993). Specifically, NGC would pay the cost of out-of-merit running due to transmission constraints (and some other factors) as captured by the uplift. NGC proposed to pay this element of the pool's costs in return for a fixed fee, levied on pool members or, alternatively, grid users. (The two are more or less synonymous.) This proposal has yet to become an official policy of NGC, but further discussions between NGC and the electricity pool are likely.

All in all, therefore, major changes are proposed for transmission tariffs in the coming years. New costing methodologies will be introduced, regional differentials will be wider, some charges may be negative, and NGC may take over a proportion of the electricity sector's costs of generation. To understand why NGC and OFFER are suggesting these changes, and whether they will succeed in fulfilling their aims, it is necessary to examine the problems that have built up since vesting in 1990.

NGC retains a monopoly over transmission, but the experience of operating a transmission system in a competitive electricity market will be of interest and concern to many electricity utilities around the world, whenever similar proposals for increasing competition in generation begin to be taken seriously.

5.4. What Went Wrong Between 1990 and 1993?

In one sense, it is possible to argue that nothing has gone wrong with the electricity transmission system in England and Wales. The lights have stayed on (weather permitting), and the market arrangements have functioned day in and day out to pay generators for supporting the system with a minimum of fuss and disruption. There has been intense public coverage of the effect of competition on prices, costs, and other energy sectors, such as coal and gas, but most of this attention has apparently concerned electricity generation and supply, not transmission.

There is, however, a growing awareness that NGC's transmission charges do not provide sufficient incentives for efficiency both to grid users and to NGC's operational departments. The following is a list of the problems that have arisen since vesting and that may gradually undermine the efficiency gains achieved through the introduction of competition in generation and supply:

1. Construction of new generators in the (far) north, where there is already a surplus of generator capacity;[3]
2. Construction of expensive, unsightly, and possibly unnecessary new transmission lines in environmentally sensitive areas (such as the North Yorkshire Moors), in order to link up the new generators in accordance with planning standards;
3. NGC's realization that the costs of such investments do not earn any additional revenue, given the revenue cap in the regulatory formula;
4. Accusations that NGC saves investment costs, by relying on out-of-merit running of expensive generators in import-constrained areas (the costs of out-of-merit generation being borne by customers of the electricity pool) (OFFER, 1992a);
5. NGC's realization that it has no incentive to make efficient investment decisions, if savings in the cost of out-of-merit generation accrue to pool members;
6. Complaints from existing generators and customers that increases in transmission charges will cause them financial hardship (especially at a time of recession) (OFFER, 1992b).

Major questions of energy policy have tended to overshadow these issues in the media, but each has been discussed in depth within the industry, and the range of possible solutions is becoming clear. To appreciate how these problems can be solved, however, it is first necessary to understand the cost structure of the transmission industry in general. The following sections therefore provides the analysis of short-run and long-run transmission costs that underlies the solutions proposed in the final section to this chapter.

5.5. The Short-Run Costs of Transmitting Energy

In a competitive market, the price of a product can never differ between two separate locations by more than the short-run marginal cost (SRMC) of transporting the product from one location to the other, including the cost of constraints. In the electricity sector, a cost-minimizing dispatcher will aim to equalize marginal generation costs throughout the system, subject to costs of transmission. In the case of the NGC, these transmission elements include transmission losses and constraints.

5.5.1. Transmission Losses

Assuming that the requirements of other users of the system remain unchanged and that there are no transmission constraints, the SRMC of any particular request to transmit electricity can be illustrated as follows.

Consider a simple two-node system (west and east). Suppose an increase in demand in the east is met at least cost by an increase in generation in the west. The net increase in the line flow from west to east will have a measurable impact on transmission losses. Any change in losses on the system is the marginal physical loss associated with the additional flow. A positive marginal loss implies that less electricity can be drawn off in the east than is supplied in the west.

Figure 5.1. A Simple Two-Node Transmission System

Figure 5.1 illustrates the case where marginal physical losses are 5 MWh; the west supplies an additional 100 MWh, but the east can only draw off 95 MWh.

The value of this loss (the SRMC of transmission) is determined by the equilibrium price difference between the west and the east. If the marginal price of generation in the west is $10 per MWh, the total cost of the additional 100 MWh is $1,000. Customers in the east will have to pay this $1,000 to cover the cost of generation even though they only receive an additional 95 MWh. The implied price in the east is $10.53 per MWh ($1,000 per 95 MWh); the SRMC of transmitting electricity west to east is therefore 53 cents per MWh.

In general, losses increase with distance, so that the farther apart the nodes are, the higher the electricity price must rise to cover the generator's marginal costs of supply. However, additional line flow raises total losses only if it moves in the same direction as the existing net flow. Any attempt to transmit electricity against the current net flow (from east to west) will reduce physical losses. More can be drawn off than is supplied and the SRMC of transmission is negative.

5.5.2. Constraints

The two main types of constraint on an electricity transmission system, thermal limits and regional voltage stability limits, usually cause the dispatcher to restrict the maximum flow over a line. This drives a further wedge between the marginal generation price either side of the constraint. However, the SRMC of transmission over the constraint is still given by the difference between these two prices.

Consider a grid system that can be divided into two zones, A and B. Ignoring marginal losses, suppose that initially the level of demand is such that the dispatcher calls up generation in both zones with a marginal cost of generation of $16 per MWh, but that this requires a substantial flow from generators in zone A toward zone B. Now impose a constraint on the amount that can flow over lines linking zone A to zone B, such that generation must be pulled back in zone A and increased in zone B. This may lead to a reduction in the marginal cost of generation in zone A to $14 per MWh whilst simultaneously raising it in zone B to $18 per MWh. The SRMC should then be $4 per MWh ($18–$14/MWh).

The SRMC of transmission can (only) be *derived* from the difference in energy spot prices between the two nodes in an optimised despatch. If the pattern of constraints is used to divide the country into zones, within which electricity can move without constraint, then:

- *Within a zone* the marginal cost is derived from the difference in price between the two nodes resulting from marginal physical losses;
- *Moving from one zone to another* the marginal cost is the value of the constraint between them.

5.5.3. Criteria for Constructing Transmission Capacity

Discussions of marginal cost pricing often get clouded by the distinction between short-run marginal cost (SRMC) and long-run marginal cost (LRMC).

Short-run marginal costs are the variable costs of production of a small increment taken now, given that almost all factors of production are fixed; or, if capacity is limited, the price necessary to ration demand to capacity. The calculation of long-run marginal costs allows all factors of production to be fully variable, and sometimes even assumes a system fully adjusted to a long-term increment.

At any instant, SRMC can be shown to be the price that would make the best efficient use of the existing resources. But this does not mean that all transactions have to be spot transactions. A three-year contract, for example, could be priced at the expected short-run marginal cost (E(SRMC)), or at the cost of marginal additions to the grid over three years, whichever is the lower.

The E(SRMC) is linked to the LRMC by the decision to construct transmission facilities. Increasing the size of a transmission link will lower the SRMC of transmission; transmission capacity should not be increased beyond the point where SRMC falls below LRMC, since it would be cheaper to tolerate a constraint.

5.5.4. Revenue Requirements and Short-Run Pricing

In the real world, revenues from marginal cost pricing (SRMC) may not equal the required revenues (LRMC). The reasons for this divergence are the good ones (economies of scale), the bad ones, (inefficiency, bad planning, and inappropriate accounting policies),[4] and the neutral ones (related to who bears any risk).

If there are economies of scale in building transmission lines, the project's earnings at prices based on SRMC may be insufficient to recoup the initial cost of the investment. In such cases, a grid company would have to levy an additional charge in order to recover the costs of the investment.[5] Such other charges may justify a deviation from prices based on strict, forward-looking marginal costs.

The various forms of inefficiency obviously need to be discouraged by ensuring that unnecessary costs are paid by the shareholders of the transmission company, rather than by grid users. The method of regulating the transmission company's resources should not, therefore, allow all costs to be passed through. For example, transmission prices might be fixed in advance under a long-term contract.

Unpredictable risks can show up as temporary surplus capacity, totally redundant capacity, or too little capacity. Whether prices should then be based on SRMC or LRMC depends on who bears the risk—the provider or the purchaser.

SRMC pricing ensures correct dispatch and is compatible with short-run incentives and efficiency: prices fall when there is excess capacity and rise to ration demand when there is a shortage. On the other hand, SRMC pricing has the disadvantage of placing a great deal of risk on any owner of a transmission network. Furthermore, although increasing prices to reflect a rising SRMC is an invitation to new entrants to enter a competitive market, a monopolist faces the opposite incentive: higher SRMC means higher prices and higher profits, so why should a monopolist build, when building reduces SRMC? These issues of risk and incentives must be addressed when regulating the provider of transmission under any system of pricing, whether prices are based on SRMC, or on some estimate of long-run costs that vary over a longer period and therefore appear to be more stable.

The current debate on transmission pricing has placed SRMC in a central position. Now that we have identified short-run marginal costs, and related them to LRMC and the revenue requirement, we are in a position to examine the proposals for pricing.

5.6. Solutions

In addressing each of its problems, NGC is drawing heavily on the economic analysis of transmission services and their (marginal) costs. These are relatively new ideas for the industry, and it will be some time before they will be sufficiently well understood to replace the current planning methods. At this point, economic analysis can only sketch out a solution for each of NGC's problems, but some of these suggestions are already being investigated, and the others deserve serious consideration. The following section indicates where the solutions will lie.

5.6.1. Generator Location: Better Signals of Marginal Cost

To encourage generators to build new plant where it is needed, NGC is trying to give stronger signals about marginal costs in its use-of-system charges. The proposed charges for the years 1993–1997 show wider (and widening) differentials between north and south, culminating in *negative* charges for generators locating on the south coast (because they actually save NGC from having to invest).

The calculation of marginal costs remains a difficult matter. Costs depend on the level of service provided, which has never been formally defined. NGC is reluctant to be committed to any particular level of service, given the variation in actual service standards that is the legacy of previous planning methods. NGC would also need a ruling from the electricity pool of England and Wales, which handles all trade in electricity, as to where the market was located, at least nominally. Without knowing the point of sale, at which energy passes from generators to customers, NGC cannot determine what proportion of total transmission costs should be borne by either party.

5.6.2. Planning Procedures: More Economic Criteria

There is a widespread feeling that the level of service to grid users could be maintained at lower cost, if the planned extension of the national grid across North Yorkshire were abandoned and replaced by alternative (and less unsightly) security measures. Unfortunately, NGC is prevented from making any such choices by a license obligation to adhere to prescribed planning standards. These documents are left over from the days of nationalization and set out the required planning procedures, regardless of the consequences for performance standards (or for the environment).

Alternative planning standards, using economic cost-benefit criteria, had actually been prepared by the late 1980s, but their implementation was shelved at privatization to avoid the accusation that standards were being lowered. Their implementation now would still require a formal definition of the level of service that NGC was required to provide to all grid users.

5.6.3. Revenue Base: Related to Level of Transmission Service

NGC's revenue is currently fixed in total during any review period (subject to some indexation for inflation). Any other company would expect its revenues to rise if it provided more services to more members of the public. Until NGC is able to define the level of service for each customer (volume and hours of transmission, from pick-up point to set-down point), it will be impossible to identify any growth (or decline) in demand for NGC's output. No regulatory formula is likely to provide appropriate incentives if it is unrelated to the level of the regulated company's output. The danger is that users will join or leave the system in ways that greatly increase or decrease NGC's costs; without a measure of each customer's usage, it will be impossible to identify the effect on costs, and revenues will not be adjusted accordingly.

Attributing costs to individual users is much simpler if each user has a contract that states what transmission services are provided to that user and includes a price related to marginal costs. As users join or leave the system, they will sign new contracts or retire existing ones, and NGC's revenues will adjust automatically to the new level of service.

5.6.4. Investment Cost Incentives: NGC Pays for Out-of-Merit Running

The marginal cost of transmitting energy over a constraint is *either* the cost of additional lines *or* the extra costs caused by running generators out-of-merit to rebalance the network. If NGC is providing transmission over a constraint, it should be required to bear the costs of doing so, whatever method is used. At present, NGC only plays the costs of additional lines; out-of-merit running is charged to customers of the electricity pool as a surcharge ("uplift") on the price of a kWh. This anomaly requires rationalization.

Either the pool should charge NGC for out-of-merit running, or NGC should charge the pool for its investments in wires. Then only one party will be responsible for both types of cost and can make rational investment decisions. Under current institutional arrangements, it makes sense to assign the costs of out-of-merit running to NGC, where the costs of wires already lie. The main obstacle to this approach is the difficulty of identifying the generation costs due to out-of-merit running, given the current pool rules.

5.6.5. NGC Incentives: NGC Pays for Out-of-Merit Running

The same solution—charging NGC for the costs of out-of-merit running— will give NGC the basis for efficient management decisions. Line outages and upgrades can both be compared with the alternative costs of out-of-merit running, to identify the investment and maintenance expenditures that are most profitable for NGC and most efficient for the system. Electricity traders are more likely to be convinced that NGC is operating the system in their interests, and NGC is more likely to be able to carry out investment without demands for detailed scrutiny. The provision of proper commercial incentives will also allow NGC to decentralize more management decisions within the company.

5.6.6. Opposition to Investment Costs: Fix NGC Charges

From 1993, NGC's total revenue will be fixed in advance until 1997, subject only to indexation for inflation. However, grid users are aware that the cost of major forecast investments was reflected in the revenue ceiling, and any additional expenditure will be recovered through extra revenue in later review periods. These fears would be allayed, if grid users knew that NGC could cover the cost of investments from savings in the cost of out-of-merit running. NGC's revenues could then continue to be fixed in advance for longer periods of time, almost regardless of the level of investment.

Of course, increasing use of the national grid might require an increase in total costs, in the form of either more out-of-merit running or additional investments in wires. In this case, the additional cost would reflect an increase in the supply of "transmission services" by NGC, which might legitimately result in increased revenues. In order to combine (1) fixed charges to existing grid users with (2) a revenue that rises in step with usage of the national grid, NGC should consider giving each grid user a long-term contract for use of the system. The price of the contract could be fixed in advance to protect existing users from cost increases during the life of their projects; new users would add to the revenues of NGC in proportion to the extra transmission costs that they impose on the system. Such contracts could even be tradable, to allow users to reassign their transmission rights to others, when their project ends and they wish to leave the system.

5.7. Will NGC Solve Its Problems?

NGC has made a brave attempt to solve the problems caused by grid users, by adjusting transmission charges (to improve marginal cost signals), while slowing the speed of adjustment to keep existing users happy. This strategy may be successful in providing good signals on future costs to both new and existing users. However, it is not clear yet whether any of NGC's proposed transmission prices truly reflect marginal costs (1) because the basis for cost calculation remains rather obscure to most observers and (2) because there is no clear definition of the level or nature of the service provided by NGC to each user. Until these deficiencies are addressed, it will be difficult to see whether NGC is providing appropriate cost signals.

NGC has made less progress on the problems caused by the incentives on NGC itself. These problems will persist until NGC receives both better incentives both externally and internally. As an external incentive, NGC needs to earn additional revenue for providing service to new users; internally, NGC should be able to trade off the costs of out-of-merit running against the cost of reinforcing the grid. Work is afoot at NGC to solve the problem of internal incentives, using the approach set out above, but the review of NGC's revenues will require a major rethink by the Office of Electricity Regulation.[6]

Provision of incentives, regulation of revenues, and protection of existing users from variation in costs would all be made easier if the current system of annual tariffs for transmission were replaced by a system of long-term contracts (or "vintaged tariffs"). So long as these contracts gave title to transmission rights that were tradable, in whole or in part, between grid users, it would be possible

- To provide NGC with exogenously fixed revenues (providing an incentive for cost minimization);
- To regulate NGC's revenues on an occasional basis, by OFFER agreeing the basis of contract terms with NGC in advance, supervising negotiations between NGC and new users, and arbitrating in cases of dispute;
- To protect existing users from fluctuations in transmission costs, if they wish to retain transmission rights, and to provide good marginal cost signals, if they would consider selling their transmission rights to other users.

Unless this major change in procedures is adopted, NGC is unlikely to be able to reconcile the conflicting pressures, will continue to face difficulties in allocating costs to users, and will become an ever greater cause of regulatory disputes.

Notes

1. The break up of the CEGB on March 31, 1990, when the new companies were created, is referred to as *vesting*. Privatization of the new companies followed in stages during 1991.

2. An exemption has recently been proposed where generation and demand are located on the same site, where generation and demand may be offset and use-of-system charges apply only to the net figure. However, the definition of a site remains unclear, and this exemption may be implemented only sporadically.

3. See NGC's seven-year statements for the years 1990 to 1992, for warnings about the level of capacity being built within northern zones.

4. In the United States, the use of a historic cost rate base completely swamps any other factors.

5. Where the total benefits (savings in costs of generation) outweigh the total costs (of construction), the investment is potentially profitable, if paid for by users in advance. If total costs exceed total benefits, the project is not worthwhile and should be rejected.

6. In November 1993, OFFER issued a consultation paper on "transmission services". In April 1994, NGC and the pool reached agreement on arrangements to share the cost of uplift over the next three years. Discussion of the relationship between uplift and transmission services is continuing.

References

National Grid Company. 1992. "Transmission Use of System Charges Review: Proposed Investment Cost Related Pricing for Use of System." National Grid Company, London, June 30.

OFFER. 1992a. "Report on Constrained-on Plant." Birmingham, October.

OFFER. 1992b. "A Statement by the Director General of Electricity Supply: NGC Transmission Use of System Charges." Birmingham, November 27.

"Power Plan Aims for Stable Prices." 1993. *Financial Times*, January 7, p. 7.

6 REGULATION OF REGIONAL ELECTRIC COMPANIES IN THE BRITISH ELECTRICITY EXPERIMENT

Michael A. Einhorn[1]

6.1. Introduction

While competition may provide feasible incentives for management efficiency in electricity generation, transmission and distribution have properties of a natural monopoly; entry has been restricted and rates have been regulated. In Chapter 5, Hunt and Shuttleworth explored inefficiencies in the regulation of Britain's National Grid Company. This chapter considers regulation of the twelve regional electric companies (RECs).

6.2. The Economics of Distribution

In March 1990, Britain's Office of Electricity Regulation (OFFER) vested twelve regional companies to operate distribution franchises. Each regional company was licensed to supply and distribute power to all customers in its franchised service territory; additionally, each may now supply power to large retail customers in other territories.[2] While distribution sales were roughly 30 percent of supply sales in the twelve regional franchises, distribution was far more profitable (OFFER, 1992c, p. 7). The RECs may price distribution using fixed tariffs or individually negotiated contracts.

Each franchised distribution grid purchases power from high-voltage transmission lines (132, 275, or 400 kV) and transports it to retail customers through a network of interconnecting overhead wires, underground cables, substations, line transformers, and consumer meters. Substation and line transformers step down line voltage at certain points in the grid; voltage step-down is performed initially at the transmission-distribution interconnection (where high transmission voltages are usually lowered to 11 or 33 kV), subsequently at locations nearer to user premises (one- and three-phase service voltages are respectively 240 and 415 V), and possibly at some intermediate locations (step-down from 33 to 11 kV is common in rural areas). Larger customers frequently take power directly off transmission or high-voltage distribution wires and provide their own step-down; this reduces distributor costs and prices charged to these large users.

In addition to installed equipment costs, voltage step-down is costly for another reason. Each electricity-using appliance or motor imposes a certain power (wattage) load on generation facilities and interconnecting T&D lines; wattage approximately is the product of current flow (amperage) and pressure (voltage).[3] Whenever electricity flows through a wire, some power is dissipated in waste heat; line losses are equal to the product of current squared and wire impedance (a physical characteristic related to length, cross-section, material, and power factor). For any wattage amount, line losses are minimized when voltage is maximized; this explains why transmission lines that transport electricity over long distances use high voltages. However, because considerably lower voltages are needed to deliver reasonably safe levels of power to end-user premises, step-down is necessary.

Seven important technical and institutional properties of electricity distribution bear upon a prospective incentive mechanism. First, distribution companies are capital-intensive entities with substantial transaction-specific costs involving long-lived assets; investments that add wire or increase voltage are irreversible, often lumpy, and often risky. Second, scale economies in transport are substantial; it makes little economic sense to build a competing grid to ensure price competition. Third, technological improvement is modest; unit costs decline primarily due to increased network throughput. Fourth, distribution companies are well suited to undertake demand-side management (DSM) programs that reduce customer loads by improving usage efficiencies and building insulation.[4]

Fifth, RECs meet most customer needs through pool purchases, although self-generation is possible. Because pool prices are volatile, RECs rely heavily on contracts-for-differences to hedge against risk. Although REC input prices are market determined and beyond any buyer's immediate control, the choice between spot and contract purchases entails some managerial strategy involving price expectations and optimal behavior under risk (see OFFER, 1992c).

Sixth, while a distribution grid entails substantial sunk costs and may be a natural monopoly, electricity supply can be competitive. Since vesting in 1990, industrial and commercial users in the United Kingdom for noncoincident peaks that exceed 1 MW have been permitted to contract with independent electricity supplies, which their local distribution franchisee must transport;[5] distribution charges are assessed to suppliers, who recover it from customers. The demand limit will decrease to 100 kW on April 1, 1994[6] and will vanish entirely on April 1, 1998. As supply and transport are unbundled, supply prices can be more competitive.

Finally, distribution engineers face tradeoffs between capital and electricity purchase costs when planning their system; new lines may reach less expensive power sources, voltage upgrades can reduce line losses, investments in demand-side management could reduce capacity and supply needs, and spot markets or contracts can be used to procure power supply. T&D arrangements in the United Kingdom have been criticized for providing underincentive for efficiency improvements (OFFER, 1992a, p. 12) and optimal risk balancing (OFFER, 1992c, p. 17).

Network regulators must then consider incentives to expand service, maintain capital investments, and procure and conserve electricity efficiently. We now consider the British experience and possible areas for reform.

6.3. Incentive Regulation of RECs

Regulatory mechanisms have been implemented in public utility markets in order to provide incentives for cost minimization and pricing efficiency. Critics of rate-of-return regulation have contended that traditional procedures provided limited incentive for managers to reduce company expenditures; when permitted rates of return exceeded market rates, utility managers may have overcapitalized rate bases (Averch and Johnson, 1962; Zajac, 1972). Second, marginal cost and second-best Ramsey pricing (Baumol and Bradford, 1970) required reliable information on company costs and user demand elasticities that is difficult to obtain. Third, intermodal (Brauetigam, 1978) or full-scale competition limited rate levels to amounts better determined through the give-and-take of competition rather than through regulatory judgment. Finally, regulators were seen as politically motivated actors who pursued personal goals in a costly adjudicatory process; the fear of regulator expropriation (such as undercompensation for sunk investments in a prudency review) may have led some utilities to underinvest in necessary plant, reversing the Averch-Johnson effect.

On April 1, 1990, OFFER implemented a price-cap mechanism for regulating the country's twelve RECs. Each REC must separately account for and recover its supply costs and the costs of installing, using, and maintaining its franchised distribution grid. If applicable, cost separation is also necessary for franchise supply, nonfranchise supply, and self-generation.

Three different price ceilings were initially instituted for each REC. Under the *overall supply price control*, per kWh charges for input costs needed to meet customer demands that arise both inside and outside of a REC's franchise territory were permitted to increase at a rate $RPI - X + Y$; RPI represents the percent change in the U.K.'s retail price index (a measure of general inflation), X represents a productivity offset, and Y represents adjustments for costs deemed beyond REC control.[7] Pass-through adjustments (i.e., Y) accounted for 95 percent of the REC total in 1992 and have included electricity purchase costs (58.3 percent), recovered distribution charges (23.8 percent), the fossil-fuel levy (9.3 percent), payments to National Grid Company (3.9 percent), and a supply business margin (4.6 percent). If company prices over- or underrecover revenues in a particular year due to unanticipated high or low demand, the index adjustment in the subsequent period is modified to make up retroactively for this discrepancy.

The designated ceiling can legally expire in March 31, 1994. OFFER (1992b, p. 18) hints that a revised control could be limited exclusively to *franchise-related* supply prices and might be based only on $RPI - X$. This is broadly consistent with the supplementary control discussed below that recently expired (see below).

This expired *supplementary supply price control* constrained supply expenses that were undertaken exclusively for franchise service. The price ceiling increased at a percent rate for increases in RPI; only changes in the fossil-fuel levy could be passed through. A dichotomy evidently existed between the overall supply price control, which was based primarily on actual costs, and the supplementary control, which was based primarily on benchmark inflation.

OFFER recently replaced the supplementary control with a nondiscrimination condition; that is, unless cost-justified, a distributor cannot charge supply prices to its franchise customers that exceed the supply prices charged to its nonfranchise customers.[8] As explained above, the overall supply price control in 1994 may come to resemble the recently voided supplementary one.

Finally, the *distribution price control* capped REC rates charged for installing, using, and maintaining distribution facilities. Average distribution charges have been permitted to increase by $RPI + X$, where X is an allowed increase to fund necessary upgrades in particular REC grids.[9] Like the overall supply price control, the index must be adjusted ex post if revenues in a particular period are too high or low due to unanticipated demand fluctuations. OFFER may eliminate this control in March 1995.

Figure 7.11 in Chapter 7 in this book illustrates inflation-adjusted prices charged to domestic customers between 1989 and 1993; price caps became operational on April 1, 1990.

6.4. Criticism

Whenever rate increases are constrained by some index other than actual cost (plus or minus X), utility managers have incentive to minimize costs. This represents the prime motivation for price caps. However, if the benchmark does not accurately track costs that are beyond utility control, the incentive scheme may be too risky for the capital-burdened utility; Schmalensee (1989) contends that low-powered rate-of-return regulation[10] may be preferable to high-powered price caps if stochastic variances and risks are great.

Knowing what costs to benchmark, freeze, or pass through requires institutional knowledge that is central to the design of a wise incentive strategy. We now explore three aspects of OFFER's incentive scheme for regulating Britain's regional electric companies: the productivity offset (X), the capital-cost adjustment, and the supply-cost adjustment.

6.4.1. Productivity Offset

The prime difficulty with price-cap incentive regulation inheres in the measure of X, which is an ex ante price adjustment chosen to represent the regulated company's expected change in unit costs. The term X is quite difficult to estimate beforehand, exposes the company to some risk, and may lead to regulatory gaming; regulators might not be able legally to permit bankruptcy and may have incentives to change X in the next regulatory hearing to appropriate profits that are deemed excessive. For example, Britain's Office of Telecommunications corrected for excessive

British Telecom profits by changing the company's annual productivity offset X from 3 percent to 4.5 percent in 1988.

A less risky way of sharing productivity gains with consumers would be to set $X = 0$ and ex post adjust prices in each rate hearing to share profits that had been realized (see Brown, Einhorn, and Vogelsang, 1989, 1991). Not requiring any estimate of X eliminates a source of risk to the utility. Profit-sharing provides financial incentives for efficient company management and provides to consumers an automatic share of the benefits that result. Chapter 13 in this book considers these issues in greater detail.

6.4.2. Capital-Cost Adjustment

OFFER's incentive mechanism does not allow regional company rates to adjust for capital costs that are incurred for network expansion, upgrade, or demand-side management; these costs are often large and nondiversifiable over time. Consequently, a utility manager who undertakes investments either to meet load growth or to economize on supply purchases may face a revenue deficiency immediately afterward. While this deficiency may be recovered over time if expected demand growth materializes, the shortfall can be unexpectedly long-lived if growth is deficient.

Since electricity purchase costs are now passed through to customers, a risk-averse manager could prefer to purchase electricity supplies instead of undertaking investments that could nonetheless reduce power needs. Such a preference may be privately rational but inefficient nonetheless.[11] This is no secret to OFFER; several parties (OFFER, 1992a, pp. 12, 15) complained that current regulation provided to utility managers no incentive to invest in equipment to reduce line losses, support demand-side management, and economize on electricity purchases.

To compensate for capital expenditures, OFFER could adjust each REC's rates for yardstick or actual capital costs. Under yardstick regulation (Baiman and Demski, 1980; Holmstrom, 1982; Shleifer, 1985), levels of or percent changes in allowed rates are based on prices or costs of comparable firms in the industry.[12] To construct yardsticks for capital costs, regulators would need to know and adjust operating and geographic characteristics of each company's distribution network; the required process seems quite complex.[13] By contrast, adjustments based on actual capital costs require considerably less human judgment; they would compensate for capital upgrades with less risk. Though pass-throughs of capital costs admittedly provide less incentive for efficiency improvements, they seem on balance to be more practical.

6.4.3. Supply-Cost Adjustment

Turning to supply purchases, an RPI adjustment is risky because overall consumer prices and supply prices can increase at significantly different rates. Furthermore, if regulators were to pass through actual capital costs (as suggested above) but adjust supply purchases for overall inflation, managers may come to view capital investment as safer and may then choose to overinvest.[14] Without passing through supply costs, OFFER might reduce unnecessary risk if it can design a yardstick that reflects a more reasonable rate of cost increase.

Cost yardsticks may be based on levels or percent changes in overall industry rates or costs. Laffont and Tirole (1993) demonstrate that stochastic shocks that would equally disturb a regional company variable and a yardstick counterpart can provide efficient incentive for manager effort and can minimize risk. By contrast, shocks that are ideosyncratic to either a REC or to its yardstick group can reward inefficient behavior and increase risk. In view of this, percent changes would seem to be a better filter of extraneous factors and would then appear to be better yardsticks. The allowed rate of increase for a REC price index could be based on a general price index or a weighted average of percent increases in input prices, categorized by provider or fuel source; weights would correspond to company cost shares. The weighted cost index would seem to control better for ideosyncratic disturbances.

6.6. Unbundling

When supply franchises terminate in 1998, every retail customer will be able to contract with any electricity supplier for power; local distribution franchisees must provide transport. At this point, supply prices presumably will be constrained by competition and beyond the need for regulation; the mechanism in Section 6.5 can be abbreviated. We now ask two questions: (1) Is unbundling efficient, and (2) if unbundling is pursued, what incentives are appropriate for the transport capital that will continue to be regulated?

Proponents of unbundling argue that it allows retail customers to negotiate directly with suppliers and permits avoidance of high-cost purchase arrangements that distribution managers may have erringly entered.[15] However, there are two objections to this proposition. First, distribution managers have less incentive under unbundling to expand and upgrade infrastructure since they cannot internalize savings in supply costs. Second, the REC's ability to dispatch facilities most efficiently is weakened if customers can independently contract for power supply and use the distribution network as a common carrier; consequently, network expansion becomes more necessary.[16]

If transport and supply were unbundled, regulators would need to provide incentives to control distributor costs, which are very capital-intensive.[17] To avoid inefficiencies in capital projects that are associated with cost pass-throughs, Laffont and Tirole (1986) and Reichelstein (1988) offer a non-Bayesian incentive scheme. Under the suggested mechanism, managers reveal a cost estimate E prior to investing; both managers and regulators would subsequently observe actual costs A. The company is compensated for actual costs plus a bonus $B = x(E) + y(E) [E - A]$, where $x(E)$ is convex and decreasing in E and $y(E) = -x'(E)$. The authors show that managers can maximize their bonus by revealing estimates as accurately as possible and by minimizing actual costs afterward regardless of prior estimated amount.

6.7. Conclusion

Great Britain's Office of Electricity Regulation and Office of Telecommunications arrested attention with their innovative reforms of the nation's utility markets. Energy planners in both developed and developing economies have taken note of the privatization, vertical disintegration, and generation arrangements now practiced in the United Kingdom. At present, the incentive mechanism that is now in place to regulate Britain's twelve distribution companies does not command primary attention. However, the problems presented by these natural monopolies may be daunting to regulators long after transitional problems in generation are resolved. Unlike telecommunications, electricity transport entails important tradeoffs between lumpy capital investments and noncapital costs; ensuring efficient incentives for capital upgrading and input substitutability should be at the center of OFFER's future initiatives.

Notes

1. Views are personal and not those of the U.S. Department of Justice. I thank David Hawkins, Martin Hall, Ingo Vogelsang, Andrew Walker, and Greg Werden for assistance. Errors are my own.

2. This power must be transported to the retail customer by the utility that is franchised to do business in the area.

3. This is exactly true for direct currents. For alternating currents, wattage is the product of voltage, amperage, and the power factor.

4. DSM theoretically is unnecessary if prices are economically efficient; this would require real-time pricing (RTP). However, RTP is not practical if the additional metering expenses needed to implement it are large. Without RTP, kWh charges during the actual peak hour would be inefficiently low; DSM may then be justified.

5. Nonfranchise providers now supply about half of all electricity sold to the 4,500 customers in this group, which accounts for 30 percent of all distributor power sold. About 1,700 of these customers purchase nonfranchise power.

6. Accounting for 50,000 customers and 50 percent of sales.

7. OFFER chose a productivity offset $X = 0$ at vesting, but this can be altered.

8. By granting franchise customers a "most-favored nation" status, OFFER may be encouraging certain distributors to compete less aggressively in nonfranchise territories in order to avoid reducing the effective price ceiling on their franchise customers (see Salop, 1986).

9. X ranges from 0 to 2.5 percent per year and is largest in the most needy grids. For example, $X = 2.5$ percent for the two RECs that serve Wales. For the more developed London REC, $X = 0$.

10. Laffont and Tirole (1993) distinguish high-power and low-power incentive structures based on how much an agent can retain of the marginal revenue product of her incremental expended effort.

11. Even if purchase allowances were benchmarked to some exogenous index, they would be more diversifiable over time than capital outlays and therefore preferable.

12. Yardstick regulation is now used for hospital payments under Medicare; reimbursement is based on average costs of treatment in similar hospitals. It also has appeared in the Spanish electricity industry and in a 1985 rate index for Illinois Power. The Federal Communications Commission recently announced plans to put price ceilings on local cable monopoly rates with prices from competitive cable markets.

13. For example, consider the regulator workload needed to implement the suggestion of Joskow and Schmalensee (1986): "Costs norms based on the statistical yardstick notion could be developed by applying econometric techniques to data on hundreds of plants and utilities, along with indices of local wages and raw material prices; such norms can be used as a basis for incentive payments."

14. Additionally, some distributors contend that RPI adjustment for electricity input prices now lead managers to buy into contracts-for-differences, which may carry a large risk premium as a result. (OFFER, 1992c, p. 17)

15. This was a major justification for unbundling of gas supply and pipeline transport that was adopted by the U.S. Federal Energy Regulatory Commission (1992).

16. When a distributor transports power from a producer node, incoming electrons move over interconnecting links in direct proportion to the relative impedances of each; physical constraints on any link may limit carrying capacity. To avoid capacity constraints, grid planners must either add capacity to a network or receive needed power at a different location.

Ideally, network use could be optimized if transport prices were based on incremental costs. This would require large metering costs, measures of opportunity costs of foregone power, and a very complex real time price structure that may not be practical or cost effective in a distribution system.

17. Operation and maintenance expenses can be regulated with a suitable benchmark for general inflation or employee wages.

References

Averch, H., and L.L. Johnson. 1962. "Behavior of the Firm under Regulatory Constraint." *American Economic Review*, 52(6): 1053–1069.

Baiman, S., and J. Demski. 1980. "Economically Optimal Performance Evaluation and Control Systems." *Journal of Accounting Research*, 18: 184–234.

Baumol, W.J., and D.F. Bradford. 1970. "Optimal Departures from Marginal Cost Pricing." *American Economic Review*, 60(3): 265–283.

Brennan, T.J. 1989. "Regulating by Capping Prices." *Journal of Regulatory Economics*, 1(2): 133–148.

Brown, L., M.A. Einhorn, and I. Vogelsang. 1989. *Incentive Regulation: A Research Report.* Washington, D.C.: Office of Economic Policy. Federal Energy Regulatory Commission.

Brown, L., M.A. Einhorn, and I. Vogelsang. 1991. "Toward Improved and Practical Incentive Regulation." *Journal of Regulatory Economics*, 3(1): 323–338.

Brown, S.J., and D.S. Sibley. 1986. *The Theory of Public Utility Pricing.* Cambridge: Cambridge University Press.

Edison Electric Institute. 1988. *Engineering and Reliability Effects of Increased Wheeling and Transmission Access.* Washington, D.C.: Edison Electric Institute.

Federal Energy Regulatory Commission. 1992. *Pipeline Service Obligations and Revisions to Regulations Governing Self-Implementing Transportation Under Part 284 of the Commission's Regulations.* Order No. 636. 18 CFR Part 284. Washington, D.C.: FERC.

Fox-Penner, P. 1990. *Electric Power Transmission and Wheeling: A Technical Primer.* Washington, D.C.: Edison Electric Institute.

Hogan, W. 1992. "Markets in Real Electric Networks Require Reactive Prices." Unpublished manuscript, Harvard University, Cambridge, Mass.

Holmstrom, B. 1982. "Moral Hazard in Teams." *Bell Journal of Economics*, 13: 324–340.

Joskow, P., and R. Schmalensee. 1986. "Incentive Regulation for Electric Utilities." *Yale Journal of Regulation*, 4: 1–49.

Laffont, J.J., and J. Tirole. 1986. "Using Cost Observations to Regulate Firms." *Journal of Political Economy*, 94: 614–641.

Laffont, J.J., and J. Tirole. 1993. *A Theory of Incentives for Procurement and Regulation.* Cambridge, Mass.: MIT Press.

Office of Electricity Regulation. 1992a. *Energy Efficiency: The Way Forward.* Birmingham, U.K.: Office of Electricity Regulation.

Office of Electricity Regulation. 1992b. *Review of Economic Purchasing.* Birmingham, U.K.: Office of Electricity Regulation.

Office of Electricity Regulation. 1992c. *The Supply Price Control Review.* Birmingham, U.K.: Office of Electricity Regulation.

Reichelstein, S. 1988. "Constructing Incentive Schemes for Government Contracts: An Application of Agency Theory." Mimeo, Graduate School of Business, Stanford University, Palo Alto, Calif.

Salop, S. 1986. "Practices that (Credibly) Facilitate Oligopoly Coordination." In J.E. Stiglitz and G.F. Mathewson (eds.), *New Developments in the Analysis of Market Structure*. Cambridge, Mass.: MIT Press.

Schmalensee, R. 1989. "Good Regulatory Regimes." *Rand Journal of Economics*, 20(3): 417–436.

Shleifer, A. 1985. "A Theory of Yardstick Competition." *Rand Journal of Economics*, 16: 319–327.

Vogelsang, I. 1989. Two-Part Tariffs as Regulatory Constraints." *Journal of Public Economics*, 39: 45–66.

Vogelsang, I. 1991. "A Non-Bayesian Incentive Mechanism Using Two-Part Tariffs." In *Price Caps and Incentive Regulation in Telecommunications*, edited by M. Einhorn. Norwell, Mass.: Kluwer.

Vogelsang, I., and J. Finsinger. 1979. "A Regulatory Adjustment Process for Optimal Pricing by Multiproduct Monopoly Firms." *Bell Journal of Economics*, 10(2): 157–171.

Zajac, E.E. 1972. "Note on 'Gold Plating' or 'Rate Base Padding.'" *Bell Journal of Economics and Management Science*, 3(1): 311–315.

7 COMPETITION, MONOPOLY AND REGULATION IN THE ELECTRICITY INDUSTRY

Stephen Littlechild

7.1. Introduction

It is now four years since the U.K. electricity industry was restructured to introduce competition in both generation and supply. All the companies in the industry—with the exception of the two nuclear companies, Nuclear Electric and Scottish Nuclear—were transferred to private ownership, and a new regulatory framework was introduced.

My duties as Regulator are set out in the Electricity Act: they include exercising my functions so that all reasonable demands for electricity are met, enabling licensees to finance their authorized activities, promoting competition in generation and supply, protecting the interests of consumers, promoting efficiency and the efficient use of electricity, promoting research and development, protecting the public from dangers, and protecting the health and safety of employees. I also have to take into account the effect on the environment and the interests of rural customers, those with disabilities, and the elderly. These wide-ranging duties may be summarized in one phrase: the protection of electricity customers and the promotion of competition.

In this chapter I hope to give some insight into how the electricity

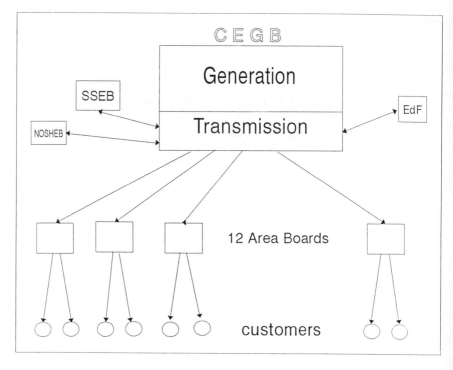

Figure 7.1. Electricity Industry Before Vesting: Structure, Physical Flows, and Customer Relationships

industry has developed over the last four years, focusing in particular on the role that regulation has played in helping to protect customers and promote competition.

7.2. Restructuring and Competition

Restructuring the industry facilitated competition in generation and supply. Figure 7.1 shows the structure of the industry as it was under nationalization, when the Central Electricity Generating Board (CEGB) was responsible for virtually all the generation plant in England and Wales, together with the national transmission grid. The CEGB supplied the twelve area boards, which in turn supplied all the customers in their area. From time to time the CEGB would trade electricity through the interconnectors from one of the two Scottish companies or from Electricité de France

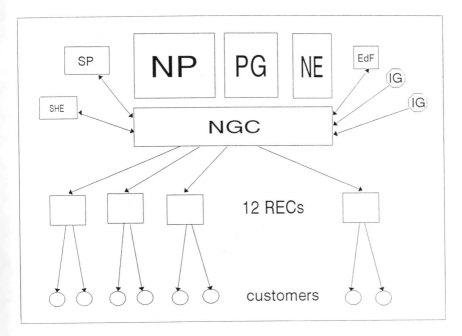

Figure 7.2. Electricity Industry After Vesting: Illustrating Physical Flows

(EdF). The figure thus shows the organizational structure of the industry, the physical flows of electricity, and the pattern of customer relationships.

Figure 7.2 shows how the industry was restructured. The CEGB was broken into four components: three generating companies (National Power, PowerGen and Nuclear Electric) and a separate transmission company, the National Grid Company (NGC). These three generation companies, plus the two Scottish companies and EdF, plus new independent generators are all able to sell into the electricity pool via the National Grid transmission system. This figure thus indicates how the structure has changed and how the generators can compete. The physical flows of electricity remain the same because the generation still flows across the national transmission system and across the wires of the local distribution systems owned by the former area boards, now known as regional electricity companies (RECs). Both the National Grid Company and the twelve regional companies have an obligation to offer terms for the use of their systems.

Figure 7.2 does not, however, adequately indicate the nature of the new contractual relationships that are now possible. These are shown as dotted

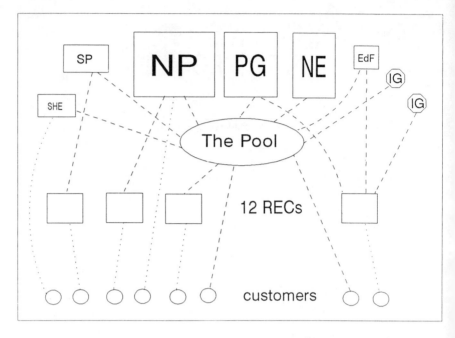

Figure 7.3. Electricity Industry After Vesting: Illustrating Physical Flows and New Contractual Relationships

lines in Figure 7.3. Generators sell into the pool, and distribution companies buy from the pool. In addition, however, any generator and distribution company can sign bilateral "contracts for differences" that, in effect, act as a hedge against fluctuations in pool price. Of crucial importance is that any generator can also sell direct to any large customer. By the same token any large customer can buy from any regional company, from any licensed supplier (all the major generators are licensed suppliers), or directly from the pool. In all cases, suppliers pay published charges for use of the transmission and distribution systems. I return to this shortly, to examine the patterns of supply that have subsequently developed.

7.3. Competition in Generation

An important aim of privatization was to encourage competition in generation. However, some have suggested that, in practice, competition in generation is a sham because the industry is dominated by the two large

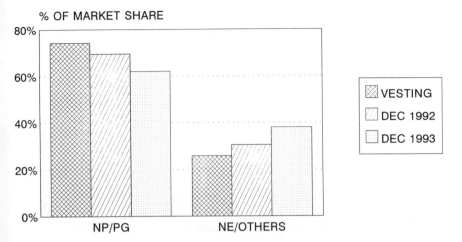

Figure 7.4. Change in Market Share of Generators Since Vesting

Note: NP = National Power; PG = PowerGen; NE = Nuclear Electric.

companies—National Power and PowerGen. To what extent is there competition in generation, and what effects has it had?

In the short term, competition must come from existing players with
existing capacity. In 1989–1990, the stations subsequently allocated to National Power and PowerGen together accounted for just under 78 percent
of the electricity sold into the pool. Nuclear Electric stations and the other
generators (principally the two Scottish companies) accounted for just over
22 percent. Figure 7.4 provides an illustration of the change in the situation by 1992–1993. Nuclear Electric and other generating companies had
expanded their share of total output in the pool from 22 percent at vesting
to just under 40 percent in December 1993. The combined share of National Power and PowerGen was correspondingly reduced from 78 percent
at vesting to just over 60 percent in 1993. It is clear, therefore, that other
existing companies are beginning to take market share from the two biggest companies.

In the longer term, it is possible for rivals to build more capacity. For
example, the Scottish companies are nearly doubling the capacity of the
interconnector so they can sell more into England and Wales. It is also
possible for new players to enter. Many new independent generating stations are being built. Figure 7.5 shows the rate at which new stations are
scheduled to come on stream. Five stations (shaded on the figure) are
being built by National Power and PowerGen. Scottish Hydro-Electric has

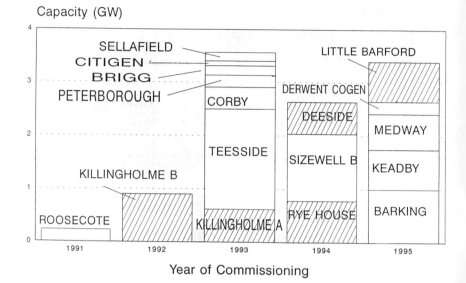

Figure 7.5.   New Generation Projects Existing and Under Construction

Note: The five shaded projects are Killingholme A and B, Deeside, Rye House, and Little Barford—all belonging to National Power or PowerGen. Nuclear Electric's Sizewell B (1254 MW) is scheduled for commissioning in 1994.

increased capacity at Peterhead, and Nuclear Electric's Sizewell B is due to be commissioned next year. The remaining eleven stations are being built by independent power producers together with regional companies. Roosecote and Enron's Teeside plant are already commissioned, another five independent stations are due to come on stream this year, and a further four stations by the end of 1995. This is a significant rate of new entry, which effectively disproves those who argued that the new industry would be too risky to stimulate new investment. It should also have a significant effect on competition in generation and on market shares, bearing in mind too that, over this period, National Power and PowerGen are implementing a substantial closure program for older, less efficient stations. As can be seen from Figure 7.6, National Power and PowerGen have, since vesting, reduced their generating capacity by approximately 8.4 GW. Further reductions of between 4 and 5 GW have been indicated by National Power over the next four to five years.

An already noticeable effect of competition is on the efficiency of the two major generators themselves. They have reappraised their investment

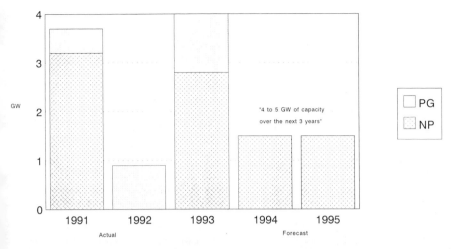

Figure 7.6.   Generation Capacity Reductions, Actual and Forecast

programs, choosing gas-fired combined-cycle gas turbines, which are quicker
to construct and more efficient to run than the large nuclear and coal-fired
plants that had been planned by the CEGB. They have also achieved
striking increases in productivity. Good progress has been achieved in
modernizing working practices at power stations. Innovative ways of running plant are being implemented to increase plant availability and thermal performance efficiencies. Output productivity in both companies
increased by about 13 percent in 1991–1992 alone. Number of employees
in both companies has been approximately halved since vesting.

7.4. Major Generators

In view of the concerns about the structure of the generation industry and
the conduct of the two major generators, I accepted the recommendation
of the former Select Committee on Energy to consider whether to make
a reference of the major generators to the Monopolies and Mergers Commission, after the market has had time to develop further but not later
than 1995. Developments in the pool and in the contracts market will be
important factors in reaching a judgment on this. So too will the development of the companies' policies on the disposal of plant. I shall therefore
be monitoring the situation particularly closely.

7.5. The Pool

The pool is a relatively novel feature of the new arrangements. Each generating company has to submit to the pool each day a declaration of what plant is available to generate in each half hour of the following day. Each company also bids a price for each of its generating sets. NGC ranks these bids in order of price and makes a forecast of what the demand will be in each half hour of the next day. It sets what is called system marginal price (SMP) for each half hour, equal to the level of the marginal bid price corresponding to the predicted level of demand in that half hour. It then adds a "capacity element" if appropriate. The result is a set of pool purchase prices.

These half hourly pool purchase prices are published each day in the *Financial Times* as shown in Table 7.1. Figure 7.7 shows these prices graphically. Figure 7.7 represents a typical pattern of prices over the day. The figure also shows the forecast pattern of demand during that day and the declared availability of generation plant. (Note: prices are given in pounds per megawatt hour (MWh). To convert to pence per unit (kWh) simply move the decimal place. Thus £20/MWh = 2p per unit).

The figure indicates how prices are higher at the times of midmorning and early evening peaks in demand, when more expensive plant has to be brought on stream. There is another peak at about 1.30 A.M. when Economy 7 heating is switched on.

The prices in Figure 7.8 are known as pool purchase price (PPP). They comprise system marginal price (SMP), which reflects the bids of generating companies, plus (as noted above) a "capacity element." This capacity element is added at times of peak demand, when available capacity is relatively scarce in relation to the level of demand. The capacity element is defined as the value of lost load (VOLL), a fixed price initially equal to £2 per kWh, multiplied by the loss of load probability (LOLP). Finally, another element known as *uplift* is added to the price at each time of day to reflect various costs of operating the system, such as payments for capacity held in reserve. The pool selling price is the sum of these three elements: SMP, the capacity element, and uplift.

Figure 7.9 shows the monthly average pattern of pool prices since vesting. It distinguishes clearly the three components just mentioned. Three significant features aroused attention and involved me as Regulator.

The first feature is the increase in capacity payments from about August 1991 to January 1992. This reflected PowerGen's strategy, when it bid into the pool, of declaring certain plant not available. This had the effect of increasing the loss of load probability and hence the amount of capacity

Table 7.1. Prices for Electricity Determined for the Purposes of the Electricity
Pooling and Settlement Arrangements in England and Wales (Financial Times)

1/2 hour period ending	Provisional Price for Trading on 19.03.93 Pool purchase price £/MWh	Final Prices for Trading on 19.02.93 Pool purchase price £/MWh	Pool selling price £/MWh
0030	18.62	18.08	18.08
0100	22.00	20.77	22.25
0130	36.60	20.77	22.24
0200	36.60	20.77	22.22
0230	18.28	18.06	18.06
0300	18.11	18.06	18.06
0330	17.98	18.03	18.03
0400	17.60	17.59	17.59
0430	17.60	17.53	17.53
0500	17.55	17.52	17.52
0530	17.64	17.60	17.60
0600	17.59	17.60	17.60
0630	18.93	18.08	18.08
0700	22.24	20.95	22.43
0730	23.51	20.82	22.28
0800	23.52	21.01	22.48
0830	24.92	21.08	22.55
0900	24.92	21.71	23.19
0930	27.13	21.71	23.19
1000	27.13	24.10	25.66
1030	27.13	24.10	25.66
1100	24.92	21.71	23.19
1130	24.92	21.71	23.19
1200	24.92	21.71	23.19
1230	24.92	21.71	23.19
1300	24.04	21.71	23.19
1330	23.52	21.54	23.02
1400	23.52	21.54	23.02
1430	23.03	21.54	23.02
1500	23.03	21.54	23.03
1530	22.95	18.62	18.62
1600	18.74	18.61	18.61
1630	18.74	18.61	18.61
1700	23.03	21.08	22.53
1730	23.03	22.46	23.94
1800	23.03	22.46	23.93
1830	28.10	23.80	25.29

Table 7.1. Cont.

	Provisional Price for Trading on 19.03.93	Final Prices for Trading on 19.02.93	
1/2 hour period ending	Pool purchase price £/MWh	Pool purchase price £/MWh	Pool selling price £/MWh
1900	28.10	23.80	25.28
1930	28.10	22.46	23.91
2000	28.10	22.46	23.92
2030	27.30	22.10	23.56
2100	21.84	22.10	23.56
2130	21.84	22.10	23.57
2200	20.56	22.10	23.57
2230	18.62	18.22	18.22
2300	17.64	18.22	18.22
2330	17.55	17.54	17.54
2400	17.64	17.54	17.54

Prices are determined for every half hour in each twenty four-hour period. Prices are in pounds per megawatt hour, rounded to two decimal places. To convert prices to pence per kilowatt hour the decimal point should be moved one place to the left—e.g., £16.86/MWh becomes 1.686p/kWh. Provision for the determination of pool prices is made in the pooling and settlement agreements which govern the operation of the electricity pool in England and Wales. The pool purchase price is the basis of the majority of payments made to generators in respect of electricity traded through the pool. The calculation of pool prices is a highly complex process the product of which is subject to revision or correction (and sometimes major alterations) until final pool prices are determined approximately twenty four days after the day of trading. Accordingly, due to the possibility of their revision and/or correction, no reliance should be placed upon provisional pool prices for any day being the same as final pool prices for that day. Final pool prices are also capable of revision. Pool selling price is the price paid by purchasers of electricity under the pool trading arrangements. It is dependent upon the determination of pool purchase price. Further information on pool prices is provided on behalf of the pool by NGC Settlements Limited. Anyone wishing to receive such information should telephone 0602–456789 between 8.30 A.M. and 5.15 P.M. Monday to Friday.

payments. The company then redeclared this capacity available on the day and earned uplift payments on it. My first pool price inquiry identified this tactic and took steps to deal with it. I agreed on appropriate license modifications with National Power, PowerGen, and Nuclear Electric. These companies are now required to provide regular information on their forecast availability of capacity, and a subsequent reconciliation with what actually occurs. Changes were also made to pool rules that removed the incentive for generators to manipulate declarations of availability in order to influence

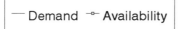

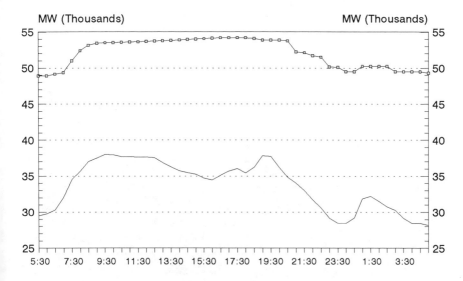

Figure 7.7. Demand and Availability for Friday, March 19, 1993

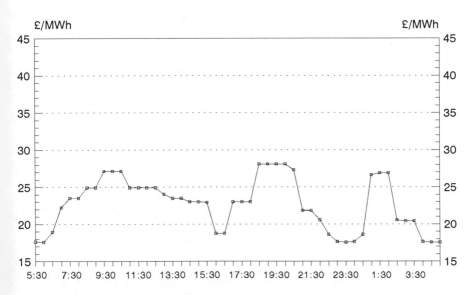

Figure 7.8. Day-Ahead Graph for Friday, March 19, 1993

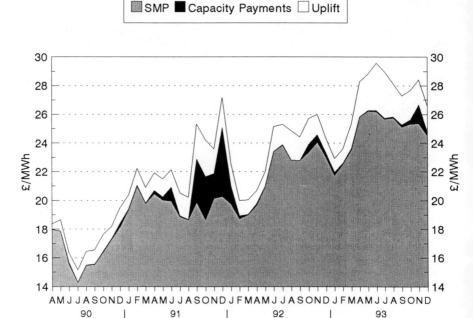

Figure 7.9. Average Pool Price by Component

capacity payments. There has been no recurrence of the practice that led to this concern.

The second feature is the increase in system marginal price in the spring of 1992. My second pool price inquiry established that this resulted from increased bid prices by the two major generators. This indicated their ability to exercise market power. However, I found that their average pool revenue was less than their avoidable cost in 1991–1992, and consequently one could not object to the increased bid prices. It is important that competitive pressures are sufficient to restrain any unjustified price increases in future.

You will notice the latest increase in pool price in April and May 1993. Customers have expressed great concern about this, and I am investigating it as a matter of priority. I have also asked National Power and PowerGen to give a public explanation and justification of their bidding behavior.

7.6. Competition in Supply

At present, only those customers with maximum demand above 1 MW can exercise the freedom of complete choice to choose their electricity supplier. There are nearly 5,000 such customers in Britain—typically large manufacturing sites that account for nearly one-third of the total electricity demand in the country. As of April 1, 1994 this freedom of choice will be extended to about 50,000 customers with maximum demand of 100 kW. This will include medium-sized businesses, supermarket chains, and many hospitals and schools. In total, the competitive market will then cover nearly half the total electricity supplied in the country. In 1998, all customers will have this freedom of choice. This ability of customers to shop around constitutes competition in supply. Whereas competition in generation is known elsewhere, competition in electricity supply was virtually unique when introduced in Britain. Although a few countries now have elements of this feature (including Chile, Norway, and prospectively New Zealand and Argentina), the extent of competition in electricity supply is greater in Britain than anywhere else in the world.

When I first raised the possibility of competition in supply, at the time of privatization, I was told that no one would be interested. Others suggested that the costs to customers of "shopping around" would outweigh any benefits. In fact, neither of these prognostications has proved correct. Competition in supply has been extremely active. Perhaps the most striking feature is that nearly one-third of the customers above 1 MW, accounting for nearly half the supply in the competitive market, have chosen suppliers other than their local regional electricity company, as illustrated in Table 7.2. The largest single supplier in the first two years was National Power, but its share is now declining. The second largest supplier was PowerGen, and PowerGen's share is increasing. The evidence is that there has been a steadily growing trend for supply outside their areas by regional companies in total, and they now account for about the same share of the total as National Power. All these trends seem to be continuing this year. There is now at least one regional company whose activities in the competitive market are greater outside its own area than they are within it. This gives some indication of the participation of a wide variety of suppliers in the competitive supply market.

Another interesting feature of this market is that about 1,000 customers take supply on terms directly related to pool prices, rather than on fixed-price contracts, although it was said they would never take such risks. The

Table 7.2. Market Shares of First- and Second-Tier Suppliers in the Nonfranchise Market

	1990–1991		1991–1992		1992–1993	
	Sites supplied	Consumption[a]	Sites supplied	Consumption	Sites supplied	Consumption[b]
REC first tier	73%	62%	65%	51%	69%	51%
National Power	12	20	14	21	8	12
PowerGen	9	11	10	15	9	18
REC second tier	5	4	10	7	13	13
Others[c]	1	3	1	6	1	6
Total nonfranchise supply	4.256	68.377 GWh	4.617	68.221 GWh	4.937	68.429 GWh

a. Consumption has been rebalanced from first- to second-tier supply by approximately 25 percent to take into account the fact that in 1990–1991, the majority of contracts applied to only nine months.

b. Forecast consumption.

c. ScottishPower, Hydro-Electric, and independent suppliers.

RECs have been active in facilitating pool-related contracts and in offering arrangements for smoothing out the variation in pool prices and for limiting the risks.

7.7. Prices in the Competitive Market

What has all this meant for the level of prices to customers? Many of the largest customers have recently expressed concern about price increases. Their concern is understandable. For many such companies, electricity represents a significant element of cost, and they are operating in internationally competitive markets at a time of recession.

The path of electricity prices has to be seen in perspective. Figure 7.10 shows government survey figures on a quarter-by-quarter basis of four different sizes of manufacturing sites. (The data for large sites were disaggregated into moderately large and extra large). Small sites would be in the franchise market, the others would be mainly or entirely in the competitive market. Pool selling price is also shown for comparison. The figures run to the end of 1992. There have been some subsequent price increases (particularly in the pool) that are not reflected here.

To make the pattern of prices easier to understand, Figure 7.11 shows them averaged over a whole year (using a four-quarter moving average). They are also adjusted for inflation by the retail price index. It now becomes evident that the introduction of competition, starting in the second quarter of 1990, led to significant real price reductions for most large customers, certainly those classified as medium and moderately large. Even despite subsequent price increases, their prices at the end of 1992 are still around 14 percent lower in real terms than in 1989–1990. For the extra large customers, however, the picture is different. Many of those who enjoyed special terms before privatization have experienced significant price increases since then. But even here the average price for the group as a whole was still about the same in real terms at the end of 1992 as in 1989–1990. As in the pool, it is of the greatest importance to ensure that effective competition keeps both costs and prices down to the minimum efficient level.

7.8. Prices in the Franchise Market

Customers who do not yet have the benefit of competition rely on regulation to protect them. Customers in the franchise market have experienced

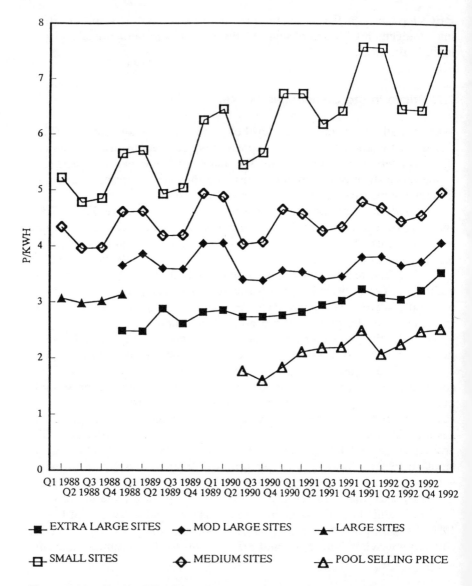

Figure 7.10. Nominal Pool and Electricity Prices to Manufacturing Sites

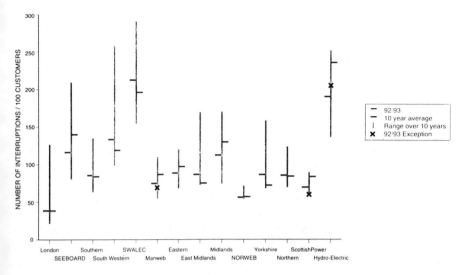

Figure 7.11.   Real (1990) Pool and Electricity Prices to Manufacturing Sites, Four-Quarter Moving Averages Deflated by the RPI

significant price increases in the past; they observe high profits earned by electricity companies and wonder if they are adequately protected.

The pattern of prices to small manufacturing sites in the franchise market is shown in Figure 7.11. Prices to domestic customers have followed a similar path but at a penny or two per unit higher. In March 1990, when the government put in place the price controls on the twelve regional companies, it explained that there would be a price increase for franchise customers in the first year—expected to be about 3 percent in real terms— but that price controls should prevent any further real increase in prices to franchise customers until April 1993. Inflation in 1990–1991 turned out to be higher than expected, and franchise customers experienced a small price decrease in real terms. The real price increase was delayed until 1991–1992. In 1992–1993, the average price increase was around 2 percent, below the inflation rate of about 3 percent, so that prices then began to fall in real terms.

This year reductions in coal prices will be reflected in lower generation costs for franchise customers. I have been able to reduce the fossil-fuel levy from 11 percent to 10 percent. I have tightened NGC's price control so that on average transmission charges will now reduce in real terms. Inflation is also lower. The price controls ensure that all these cost reductions are passed on to customers. Some regional companies have announced

Table 7.3.   Timetable of Price Control Revision

1993	National Grid Company
1994	Scots transmission REC supply business
1995	Scots distribution and supply REC distribution business

price freezes for this year, others have announced reductions. In real terms, all companies are reducing their prices.

In the future, domestic electricity prices will be subject to VAT. The possible effect of environmental measures such as carbon taxes or limits on sulphur emissions could raise costs. On the other hand, one can envisage reductions in coal prices, the levy, and transmission charges. The price controls presently under review will ensure that customers benefit from such cost reductions and will bring greater pressure to bear on supply margins and distribution charges.

7.9. Price Control Reviews

Let us now look at the process of reviewing price controls. When the regulatory regime was put in place, price controls were embodied in the licenses of the National Grid Company (NGC), the twelve regional electricity companies and the two Scottish companies, together with a timetable for revising these controls (see Table 7.3).

The first review was of NGC's transmission business, which has now been completed. In revising NGC's price control, my aim was to keep its prices down to the minimum necessary to sustain an efficient business, while leaving NGC with an incentive to improve efficiency still further. In July last year, after a thorough examination of NGC's costs, I proposed a new and tighter price control, which the company accepted. This came into effect on April 1, 1993. The revised control imposes a reduction in transmission charges of 3 percent a year in real terms.

There are two price controls affecting the revenues of the regional electricity companies (RECs)—one on supply and one on distribution. Revised controls are scheduled for implementation in 1994 and 1995, respectively. Work was started in the middle of 1992 on the review of the supply price control, which covers electricity purchase costs and the margin taken by the RECs to cover direct selling costs such as billing. I expect to make proposals for revised supply price controls by the summer of 1993. This allows time for a reference to the Monopolies and Mergers Commission, should any company decide not to accept my proposal.

While discussions with the companies continue, it would not be appi priate for me now to predict the outcome of this review. However, I intend that the revised supply price control should give firmer protection to customers who remain in the monopoly franchise market, while leaving RECs a better incentive to compete in the nonfranchise market.

Work is also under way to review the REC distribution price controls. Some people, observing high profits earned by the regional companies, have suggested that I should bring forward this review. While I understand their argument, I am not convinced that a change to the timetable laid down in the license would in fact be helpful to customers. It would introduce additional uncertainty about regulation. This would increase the companies' cost of capital and reduce their incentive to cut other costs. Ultimately, higher costs mean higher prices to customers. I believe that these higher costs and prices, which would continue into the future, would more than offset any short-term gain from lower prices for a year or two.

I therefore believe that the best interests of customers will be served by continuing with the price control review according to the scheduled time-table. Of course, the levels of the revised price controls will need to reflect appropriate levels of return to capital relative to risk, with no assumption that the present levels of profit are appropriate for the future.

7.10. Standards of Service

Customers are, of course, interested in quality of service as well as price. It is no good regulating prices if quality of service is allowed to deteriorate. Regulation therefore provides for this. In the first place, I monitor and publish information about a wide variety of technical services standards. For example, Figure 7.12 shows for each public electricity supplier the number of supply interruptions per 100 customers. These interruptions can be caused by many different factors, such as faults or damage in the network, the effects of weather, prearranged outages, and so on. The slide shows the range of performance over the last ten years, and the performance in 1992–1993 compared to the ten-year average.

Manweb, ScottishPower, and Hydro-Electric indicated that they were subjected to extreme weather conditions during the year. The graph includes an additional data point (1992–1993 exception), which shows these companies' performances when those extreme conditions are excluded. Despite this, many companies show better-than-average results, and even the results given by those companies that claimed extreme weather conditions do not exhibit severe irregularities. In addition to the three companies mentioned

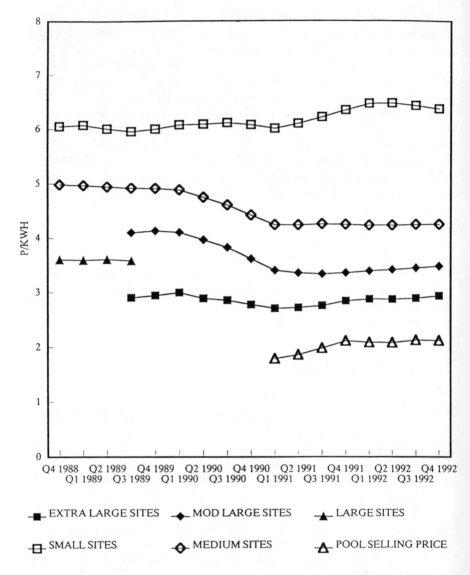

P/KWH

Q4 1988 · Q2 1989 · Q4 1989 · Q2 1990 · Q4 1990 · Q2 1991 · Q4 1991 · Q2 1992 · Q4 1992
　Q1 1989 · Q3 1989 · Q1 1990 · Q3 1990 · Q1 1991 · Q3 1991 · Q1 1992 · Q3 1992

■ EXTRA LARGE SITES   ◆ MOD LARGE SITES   ▲ LARGE SITES

☐ SMALL SITES   ◇ MEDIUM SITES   △ POOL SELLING PRICE

Figure 7.12.   Security Supply Interruptions per 100 Connected Customers

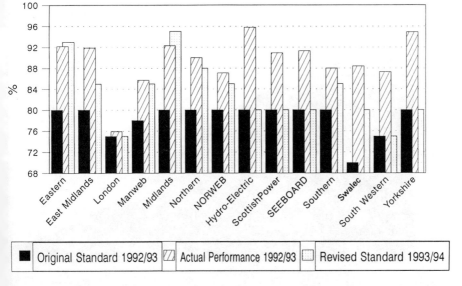

Figure 7.13. Overall Standard 1(A): Reconnect Supply Within Three Hours, Original and Revised Standards and Actual Performance

above, Eastern Electricity referred to the effect of the extraordinary high level of lightning activity in its region during the year, which caused 1,150 HV faults compared with 330 in the previous year; 550,000 customers were affected by these faults (127,000 in 1991–1992).

I also have power to prescribe minimum standards of service, and I exercised this power shortly after vesting. Last July, I published the first of what will be an annual series of statistics showing how companies performed against the standards of performance which I set. Figure 7.13 illustrates with one example—the percentage of supply failures that are reconnected within three hours. For most companies I set a standard of 80 percent, and all companies beat the standard set for them. Eight companies have volunteered to meet higher standards this year, and I have revised their standards accordingly.

In general, I am pleased at what the companies have achieved here. There is every reason to believe that standards have been maintained, and in many cases improved, since privatization. Furthermore, comparison between companies shows what best practice can achieve. I hope that the statistics published later in 1993 will demonstrate further real improvements in service to customers.

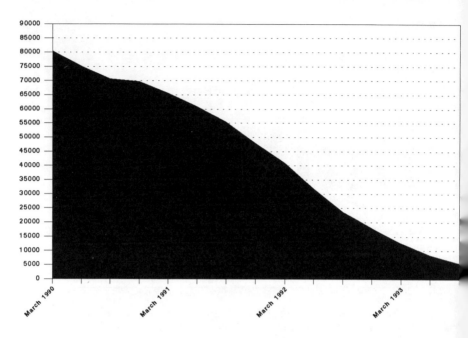

Figure 7.14. Disconnection of Domestic Customers During Preceding Twelve Months in Britain

Another innovation is that companies must make compensation payments for failure to meet guaranteed minimum standards. From April, I have doubled from £10 to £20 the general level of these payments. I am also enhancing the standard under which companies must make and keep appointments with customers. Where a customer so requests, companies must now offer and agree to an appointment within a specified two hour time band, and must make a compensation payment if they fail to meet it.

I have consistently made clear that I expect companies to reduce the number of customers whose electricity supplies are disconnected. As Figure 7.14 shows, companies have continued to make progress here. Disconnections in England and Wales are now some 75 percent lower than before privatization, and in Scotland 82 percent lower.

What do customers themselves think? We have carried out a survey using MORI that reveals some of their views, some of their criticisms, and some of the things they think have improved. Figure 7.15 shows the areas where customers told the MORI survey that things had got worse. About

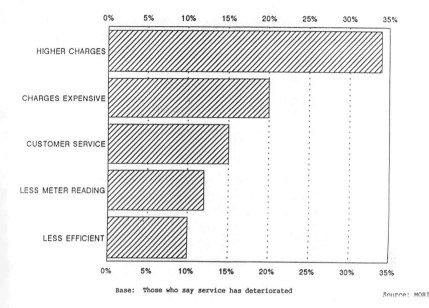

Figure 7.15.   Perceived Deteriorations in Electricity Company Services

8 percent of customers felt service had deteriorated since privatization. Their biggest single complaint was about higher charges, with about one-third of them concerned about this. About 20 percent of them said that charges were expensive. There were also some complaints about customer service and about less frequent meter reading, and some thought that the companies were less efficient. The main criticism, however, has been of prices.

Figure 7.16 shows the more satisfactory aspects of the change. About 7 percent of customers saw positive overall changes since privatization. About a quarter of them perceived prompter replies to their complaints, 14 percent claimed more efficiency, 12 percent believed that companies come out more quickly when called, 12 percent saw an improvement in customer service, and 10 percent noted better and more choices of ways to pay.

I have mentioned that 8 percent of customers felt service had deteriorated since privatization and 7 percent thought it had improved, but what about the overall situation? We asked the question, "What is your overall degree of satisfaction with the electricity service you now have?" and 6 percent of customers said they were dissatisfied while 45 percent said they

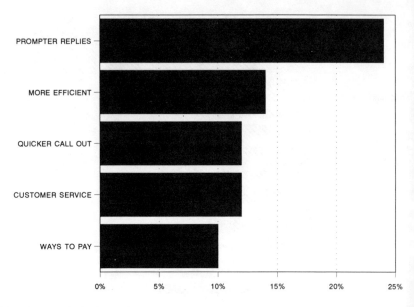

Base: Those who say service has improved

Figure 7.16. Perceived Improvements in Electricity Company Services

were either totally satisfied or very satisfied. Thus, there appears to be a broadly satisfied customer base.

7.11. Conclusions

There have been encouraging developments in the electricity industry. Certainly there have been concerns about various aspects of the new arrangements, particularly about the extent and effectiveness of competition in generation, especially in the pool, and I shall be monitoring this closely and taking whatever action is necessary. In this chapter, however, I have tried to explain how customers are actually being protected in other respects. The main points may be summarized as follows:

- There is evidence of increasing competition in generation.
- Most large customers have enjoyed significantly lower prices in real terms than before privatization.

- Prices to franchise customers have been held to the levels prescribed in the price controls, and prices are falling to reflect cost reductions in 1993.
- Standards of service have been maintained or improved since privatization, and the number of disconnections continues to decline.

Further improvements can and will be sought throughout the industry. As Regulator, I shall play my full part in ensuring that the new arrangements deliver benefits to customers.

8 PRACTICAL CONSIDERATIONS IN RESTRUCTURING OF ELECTRICITY SUPPLY INDUSTRIES

James V. Barker, Jr.
William H. Dunn, Jr.
Mk P. Shean

8.1. Introduction

The industry sector that is responsible for the supply of electricity is undergoing evolutionary and revolutionary restructure. The introduction of all source competitive bidding for generation in the United States, restructure of the entire industry in a major supply region within Chile, and removal of an obligation on any entity to install capacity to meet demand in England and Wales, have served as major milestones in a trend that is gaining momentum in the early 1990s.

The value of an entire industry, or of parts of the industry, is largely determined by its structure and the structure and stability of state institutions that regulate the industry. This chapter is intended to provide a practical framework for considering and evaluating options for structuring an electricity supply industry.

Though much of what is discussed here may be regarded as basic and not on the cutting edge of theory, satisfactory resolution of these practical issues is fundamental to successful restructuring and, where necessary or desired, privatization. Development of conceptual approaches should be based on a knowledge of real-world technical and commercial constraints,

not classroom theory. It is our experience that even the most elegant
algorithms and models fail to capture the wide range of behavior that
characterizes competitive markets.

In order to develop a structure which is best situated to permit achievement of the goals for restructure, it is necessary to recognize and understand six major components of electricity supply:

1. Generation (production),
2. Transmission (delivery of bulk power, a wires business),
3. Distribution (delivery to wholesale and retail customers, a wires
   business),
4. System planning and system operations,
5. Bulk power markets (sales at wholesale), and
6. Sales (energy sales to retail customers)

Therefore, the following discussion reviews some of the varied objectives
for restructuring and each of the major components. Alternative electricity
supply structures are then reviewed through reference to three countries
that illustrate options for allocating the six major functions among the
privately owned or state-owned entities. Finally, eight significant implementation considerations are identified and discussed.

The structure of the bulk power market (power contracting), market
mechanisms (broker or pooling arrangements), transmission pricing, and
regulation are major issues for consideration in restructuring electricity
supply, but a full discussion of these subjects is beyond the limited scope
of this chapter.

8.2. Objectives for Restructuring

What is the reason for restructuring electricity supply? Do real or perceived problems justify radical or incremental changes? There is always a
danger that ideology and dogma will dominate practicality and common
sense. Therefore, before discussing the options that may be considered in
restructuring an electricity supply industry, one must consider the purpose
for altering an industry that, especially in developed countries, has generally established an admirable record of providing reliable high-quality
electricity service at a reasonable price.

In each country or region of a country, unique circumstances must be
considered in developing realistic objectives and in designing structures
that have a reasonable chance of achieving those objectives. For instance,

the objective of obtaining needed capital for a developing country is quite different from the intent to introduce greater efficiency into a sophisticated, but perhaps bureaucratic, industry.

Views of appropriate objectives will differ as a function of the perspective of individuals. Government officials may perceive inefficiencies that harm consumers and competitiveness in international markets. Industry officials may believe that they manage efficient operations that do not require any change. However, other utility officials may perceive opportunities for reduced investment costs and a reduction in risk to shareholders from alternative methods for meeting generation expansion requirements. In Eastern Europe and in the former Soviet Union, emerging nations may wish to attain a level of electric energy independence from one another.

In general, it may be observed that those objectives that are intended to achieve quick results create a greater risk for structural design errors than do objectives that permit a more considered approach. This is particularly true if multiple objectives are attempted at the same time—such as restructure *and* privatization. Also, objectives that are intended to achieve political dogma may introduce a risk of expedient decisions taken to meet political considerations without regard to real-world complexities, thereby creating additional problems as the restructuring process unfolds. The latter has been a characteristic of restructure in England and Wales as evidenced by the twists and turns of a government that seems unable to decide the right balance between encouraging increased competition, preserving its coal and nuclear industries, and allocating costs among consumer classes. Within the first two and a half years after privatization in England and Wales, two select committee investigations have resulted in significant recommendations for fundamental changes.

The emphasis in this chapter is on *restructuring*, rather than *privatization*. The first-order problem is to create the most efficient relationship among the major electricity supply functions. Once this is accomplished, a second stage may be to privatize state-owned sectors of the industry. A change in ownership will not necessarily correct structural problems: there is no assurance that a privately owned electric utility will be any more efficient than a state-owned utility, unless a greater degree of effective regulation, competition, or a combination of both is introduced. There is no reason why either state- or privately owned monopolies should be exempt from regulation. However, state-owned agencies may have broader objectives than privately-owned institutions, such as achieving social and economic objectives. State-owned utilities that regard themselves as public works entities, for the purpose of providing employment, are not likely to be as efficient as regulated privately owned utilities that have a more narrow focus on low cost, reliable electric energy supply.

In the case of the United States, where a major part of the industry is already privately owned, the principal concern is restructure. In some instances, restructure is being driven from within by utility executives who are concerned about increasing efficiency and competitiveness. In other instances, restructure is resulting from federal and state pressure to open the bulk power market to more players. The 1992 Energy Policy Act will serve to increase such pressures.

In developed and lesser developed countries where electricity supply has been a state-owned service, the issue may be both restructure and privatization. However, the *structure* of the industry to be privatized, in such cases, should be considered first, implemented, and, if possible, proven before attempting to sell parts of the industry to investors. As the conflicts and debate in England and Wales following privatization of its electricity supply industry, show, it is much more difficult for a government to make structural adjustments after an industry has been sold than before.

The critical need for capital in lesser developed countries may make it very difficult to delay privatization, at least for new facilities, until a new structure and institutions can be established and tested. However, developed countries should be able to proceed at a more gradual pace, thereby reducing the risk of selling state-owned assets for less than an optimal market value.

The objectives, increased efficiency or acquisition of capital, should be clearly identified and referred to continually throughout the process of restructure. Conflicting objectives—promoting increased efficiency *and* preservation of high-cost domestic fuel industries—should be identified and resolved *before* privatization, not after.

8.3. Major Electricity Supply Functions

As indicated earlier, electricity supply may be viewed as a puzzle with six pieces. As with any attempt to encapsulate and describe the industry, even these six functions represent an oversimplification, it being possible to further subdivide each of these functions. However, we suggest that these are the six major components that have been and are being grouped in a variety of ways. Therefore, we discuss each of the six and alternative methods for combining the functions.

8.3.1. Generation

The generation function generally includes planning, designing, constructing, procuring fuel for, and operating the facilities necessary to produce electric energy for delivery to consumers.

Generation has rapidly come to be viewed as an excellent candidate for separation from the traditional vertically integrated industry structure. In one region of Chile, and in England and Wales, most of the existing generation has been sold to private investors. Entire plants, or shares of generating plants, are being sold throughout the world.

In the United States, the amount of generation provided by nonutility sources was greatly stimulated by the Public Utility Regulatory Policies Act of 1978 (PURPA), which created substantial incentives for other industries to build their own generation to meet their own requirements and to sell excess electric energy to utilities. While this scheme may have resulted in high costs to consumers, due to the excessively high, administratively determined, avoided cost rates being paid by the purchasing utilities, it has clearly also established the viability of integrating generation projects developed by entities without retail customers into the overall electricity supply industry.

The United States electricity supply industry has developed the term *nonutility generation* (NUG) to describe such generation. The term *utility* may require redefining in a world in which electricity supply is deintegrated. For the sake of convention, we will use the term NUG in this discussion.

8.3.2. Transmission

The transmission function generally includes planning, designing, constructing, and operating the facilities necessary to deliver bulk energy from generating stations and from neighboring systems to terminals at which the voltage is transformed to lower levels for delivery of energy to customers. Some customers are, however, connected at transmission voltage levels.

It continues to be accepted that the most economic method for providing the transmission function is in the form of a monopoly business; that is, that construction of parallel transmission lines by competing entities would result in higher costs than could be attained by having a single entity to provide such service. However, a corollary to meeting this functional requirement, as a monopoly, is that there should be some form of regulatory oversight to mitigate against the abuse of monopoly power.

While the methods for pricing transmission are beyond the scope of this chapter, it is important to note that the pricing for access to and use of the transmission system needs to support the bulk power market structure, not dictate it. It would not be logical for transmission, which usually represents a minor percentage of the total cost of electricity, to limit the bulk power market, which constitutes a much greater portion of the cost of electricity. It would also not be logical to have an unstable transmission pricing mechanism that could both send confusing signals to users and place the investment in the transmission system at risk.

While access to transmission service—third-party access (TPA) in Europe—has been much debated, creation of transmission-only utilities is a relatively recent development. In India, the Power Grid Corporation (formerly the National Power Transmission Corporation) was created in 1989. In England and Wales, the National Grid Company was created as a transmission entity, with additional separate business functions, in 1990. In 1992, Statnett in Norway and Svenska Kraftnät in Sweden were also created as separate transmission-only entities. In New Zealand, TransPower Ltd. provides transmission service. In the United States, a transmission company has existed for more than thirty years in Vermont, the Vermont Electric Power Company.

Transmission is a *wires* business and should not be confused with *generation control*, even though some transmission entities may also be responsible for generation scheduling and dispatch. This is discussed further later in this chapter. Also, transmission losses and treatment of transmission constraints must be addressed in the development of the industry structure and not be left as an afterthought.

8.3.3. Distribution

The distribution function provides for delivery of energy to the end-use customer. The distribution function includes planning, designing, installing, and operating the facilities necessary for delivery of energy at lower, as compared to transmission, voltage levels. The distribution function is similar in many respects to the transmission function. It is also a *wires* business. However, the distribution system is much more closely related to the end-use consumers, and it is typically more vulnerable to outages as a result of accidents or storms.

The distribution wires business, similar to the transmission wires business, is regarded as being a natural monopoly and, therefore, should be subject to some form of regulation in order to protect customers from a utility exercising monopoly power. Today there are very few, if any, distribution wires only utilities. Most generally, the distribution wires function is combined with the sales function. However, in England and Wales the possibility exists, after 1998, for the distribution utilities—regional electricity companies (RECs)—to be left with only their wires businesses. This could only happen if other entities are totally successful in competing to sell energy to the distribution utilities' prior customers.

8.3.4. System Planing and Operations

The system functions' scope of concern (planning and operations) encompasses both aspects of the bulk power system (generation and transmission).

System planning is the process of identifying the need for the new facilities or demand-side management (DSM) that is required to accommodate demand and energy growth. System operations is the process of planning, scheduling, and dispatching the resources, both supply-side and demand-side, to meet reliability and economy objective functions. The system functions are intended to ensure both a minimum cost expansion of the system, considering capital and operating cost of generation, transmission and DSM, and a minimum cost reliable operation of the system, considering incremental costs.

Long-range planning may be performed independently by the generation and by the transmission functions, if they exist in two or more entities. However, a check is needed to protect consumers from development of plans that are financially favorable to a single entity while being detrimental, in terms of higher long-term costs, to the ultimate consumer of services provided by the entity.

Once investments in electric generating plant, DSM, and transmission and distribution facilities have been completed; it is the function of system operations to schedule and dispatch all available resources with the objective of providing reliable electricity to consumers at the lowest marginal operating price. This objective is based on the assumption that the retail tariff structure recovers investment costs.

The system operators, generation schedulers and dispatchers possess the most in-depth understanding of the alternative variable costs (prices) of various sources of generation. They must also know the strengths and the weaknesses in the transmission network, which may require dispatching higher cost (price) generation at certain nodes (substations) in order to ensure adequate protection against emergencies and in order to meet frequency and voltage requirements.

8.3.5. Bulk Power Markets

The bulk power markets function is the process by which the buying and selling of bulk power products occurs among trading entities. The buyer in bulk power markets is not the final customer, except in the case of large customers. Typically, the bulk power buying entity resells electricity to the final customer, the end user.

It is important to note that the bulk power market's function must coordinate closely with the system operations function. The system operations function initiates the generation control actions that are required to implement delivery of a bulk power product from one trading partner to another. Therefore, system operations considerations act as a constraint on the bulk power markets and should be understood in order to avoid negotiating transactions that will either cause operational problems or which must be ignored in order to meet system reliability requirements.

Most bulk power transactions are bilateral where each of the entities is a vertically integrated utility or where the buyer is a distribution utility that, in turn, resells to the final customers. In a limited number of cases, the seller or the buyer may be a power pool or a broker.

The basic products that are traded are capacity and energy. Capacity sales usually entitle the purchaser to energy purchases at an agreed price. If capacity transactions do not include must-take energy provisions, they may be viewed as option contracts (that is, the energy is taken only when economic). These have been widely used in the United States. In instances where alternative energy may be purchased for less than the energy price agreed in the capacity and energy transaction, a purchaser of energy may elect to purchase the energy from the lower price source. In other capacity and energy transactions, there may be a must-take energy requirement. The combinations for packaging these two basic elements, capacity and energy, are limited only by the imagination of the trading parties and the regulatory regime(s) which may have jurisdiction.

The market products include a number of services that have traditionally been bundled and that are, increasingly, being viewed as separate services that may be sold independently or as parts of a package. Some of the products include:

- Load following service
- Frequency regulation service
- Spinning reserves
- Quick-start reserves
- Reactive power capacity and energy
- Black-start capability
- Scheduling and dispatch service
- Transmission or distribution losses

In England and Wales, the first six of these products are characterized as ancillary services and are bought by a business unit of the National Grid Company (NGC). The seventh product, scheduling and dispatch, is blended with NGC's wires business and is included in its charge for connection to and use of the transmission system. The eighth product, transmission losses, is accounted for in the pool settlement system by allocation to retail suppliers.

The electricity products may be traded through various market mechanisms, including forms of borkering, power pooling arrangements, bilateral transactions between the producer and the buyer, and bilateral transactions between buyers. It is not unusual for multiple market mechanisms to operate in parallel, either for the same or for different products.

8.3.6. Retail Sales

Retail sales is the function of selling electricity to the final customer, typically at a single tariff rate that includes the full range of services that are required to provide reliable electric energy.

The tariff may be differentiated among customer classes in order to reflect the cost of service to each class or to provide subsidies. Geographical and class cross-subsidation have proven to create an especially difficult problem to resolve when introducing a deintegrated electricity supply industry structure. The tariff may also be designed to provide incentives—perhaps to reduce peak time usage in order to avoid construction of new generating capacity.

The introduction of retail competition has elevated sales of generating capacity and energy as a major function that must be considered independent from the distribution wires function. England and Wales are on the road toward total retail competition in 1998.

8.4. Alternative Electricity Supply Structures

Restructure of electricity supply industries in the 1980s and early 1990s has provided tangible examples of alternative arrangements of the six major functions. However, because of differing political, economic, and cultural circumstances, the restructuring has occurred in widely different manners: the rapid and dramatic changes in the United Kingdom contrast with evolving changes in the United States.

The United States, with more than 3,000 electric utilities, offers a variety of examples or arrangements of the six major functions. The largest amount of sales in the United States are by vertically integrated utilities that include all six major functions. However, there are also many distribution utilities that purchase electricity at wholesale for resale to retail consumers. Private generators are an increasing component in the supply of electricity to vertically integrated utilities. The United States also has several federal- and state-owned generation, transmission, and power marketing agencies. These utilities also include the system planning and operations functions and sell bulk electricity to both vertically integrated and distribution utilities. As of 1993, there is one pure transmission company. In the mid-Atlantic and northeastern states, the system operations function is divided hierarchically between power pool dispatch operations and subpool or member utility dispatch operations.

There has been an evolution in the electricity supply industry within the United States and other parts of the world to implement the functions described above within existing corporate entities, but along functional lines. For instance, the Public Service Electric & Gas Company in New Jersey has reorganized into business units.

In Australia, the State Electricity Commission of Victoria established strategic business units for generation, transmission, and distribution and sales combined. The system planning and system operations functions reported to the head of the utility in parallel with the three strategic business units. Initially, the procurement of power from the production function was performed by the system operations function. However, this procurement has now shifted to between the production function and the distribution and sales function. These business units next became seperate corporate entities.

These business unit reorganizations, and others that have been instituted among utilities within the United States, have been intended to reduce bureaucratic inefficiencies that have developed in noncompetitive environments. An executive is assigned responsibility and is held accountable for the performance of a business unit. By breaking a large entity into smaller parts, problems may be more easily identified and fixed. The establishment of such business units also facilitates the possibility of further spin off of these business units into totally separate corporate entities and, for state-owned utilities, facilitates the process of privatization, e.g. Victoria.

Several illustrations have been noted in discussing each of the independent utility functions. Much has been published with respect to England and Wales. In England and Wales, the electricity supply industry divides the major functions into five levels:

- Generation,
- Transmission and system operations and transmission system planning,
- Distribution,
- Bulk power market, and
- Sales.

The bulk power system operation and transmission planning functions are performed by the transmission company. Affiliates of the generating companies, supply businesses of the distribution companies, and independent suppliers are required to purchase from a power pool for resale to retail customers. As of 1998, every retail customer will be permitted to purchase from any licensed supplier. A subsidiary of the transmission company administers the settlement and accounting for the power pool. All sales, by generating units above 10 MW, are to the power pool. All purchases by licensed suppliers, above 0.5 MW, are from the power pool. A market in hedging contracts between generators and suppliers stabilizes the pool prices. Proposals for bilateral trading *outside the pool* may lead to establishment of increased choices among trading partners. There is no requirement for any entity to install generating capacity.

Other examples are provided in the following discussion to illustrate other structural arrangements. These examples include Northern Ireland, the Netherlands, and Norway.

8.4.1. Northern Ireland

In Northern Ireland, the six functions are divided into four levels:

- Generation,
- System planning and operations and the bulk power market,
- Transmission and distribution, and
- Sales.

Restructuring has been completed and all functions have been privatized. Further restructure in bulk and retail sales is under consideration. However, initially each of the generating plants has been sold and power purchase agreements (PPAs) are in effect between each of the new generating companies and the power procurement function within Northern Ireland Electricity (NIE).

The NIE power procurement manager is responsible for bulk power purchases, system planning, and system operations. The power procurement manager dispatches generation, purchases energy from all sources, and then resells it to the affiliated NIE public electricity supply (PES) business or to other retail sales entities (second-tier suppliers) at a bulk supply tariff.

The transmission and distribution functions are also provided by the NIE PES entity, but no cross-subsidation is permitted between the delivery business and the sales business. It is possible, under the initial structure, for creation of multiple competitors for retail sales to retail customers. However, each competitor is required to purchase energy from the power procurement manager at the bulk supply tariff.

The NIE power procurement manager and the NIE PES were sold as a single corporate entity in 1993. The licenses under which this new structure was established permit possible future implementation of direct contracts, beginning in 1996, between second-tier suppliers and the generating companies. This would result in wholesale competition and could replace the bulk supply tariff.

8.4.2. The Netherlands

A second illustration of division of the functions exists in the Netherlands. The major functions are divided among three levels:

- Generation,
- System planning and operations, the bulk power market, and transmission, and
- Distribution and sales.

In the Netherlands, four separate regional production companies, each of which owns and operates multiple generating plants, are responsible for bulk power sales to the Dutch power pool (SEP).

SEP includes the system planning and operations, bulk power market, and transmission system functions. SEP is responsible for planning for generation additions and contracts with the regional production companies to build new generation. SEP directs the commitment and dispatch of generation and purchases the power from the generating facilities. It then bundles energy from all sources and sells that bulk power at a uniform tariff, including all transmission and pool administrative costs, back to the regional production companies. SEP also owns and operates the transmission system.

The regional production companies sell energy to distribution utilities at regional bulk supply tariffs that add overhead and fixed costs to the national bulk supply tariff. The distribution utilities provide the delivery and retail sales functions. The distribution companies are permitted to shop for energy from sources other than their regional production companies but, as of the early 1990s, none had yet done so.

All of these entities are regionally or federally owned. The industry is not owned by private investors.

8.4.3. Norway

In contrast to Northern Ireland and the Netherlands, Norway primarily consists of many vertically integrated utilities. The major functions may be viewed as being divided into three levels:

- Generation, system planning and operations, regional transmission, distribution, and sales,
- Pool bulk power market, coordination of system operations, and transmission, and
- Bilateral bulk power market.

In Norway, the generation and system operations functions are dispersed among a number of entities, most of which also include the distribution and sales functions. The thirty-three largest power producers own 96 percent of the production capability. There are currently over 200 different companies supplying energy to customers in Norway.

The transmission function is dispersed among the national transmission company and regional and municipal companies. Multiple bulk power markets operate in parallel. Retail sales are the responsibility of the distribution utilities, many of which are integrated with or which are owned by producing entities. Some power producers own multiple companies supplying energy to customers.

Beginning in 1991, retail competition was considerably expanded in Norway. In 1992, a transmission company, Statnett, was created with the requirement to provide nondiscriminatory access to the transmission network by both producing utilities and consumers, the consumers being primarily local or regional distribution utilities. The market had previously been conducted at two levels, one being bilateral transactions between producers and buyers, and the second being through a market mechanism established by the Norwegian power pool.

The Norwegian power pool also provided operational coordination among the utilities and instructed generation loadings for network security reasons. Generators were compensated, according to pool rules, for adjustments that were required to their schedules.

Following the establishment in 1992 of Statnett, the state-owned entity that now provides the transmission function, it was decided to merge the Norwegian power pool into the transmission entity in 1993 as a separate business unit called Statnett Market Limited. The bilateral transactions, which account for eighty percent of the total sales, continue to exist side by side with transactions through Statnett Market.

While there are many vertically integrated utilities, the largest producing utility is Statkraft, which formerly provided the transmission service and which does not sell at retail. Most of the producing utilities operate their own control centers. Therefore, the system operations function is disbursed among a number of entities.

In the new structure, producers provide reserve capacity according to common rules. The four interconnected NORDEL countries (Denmark, Finland, Norway, and Sweden) also have a common operating reserve policy.

Independent trading entities have emerged, which purchase energy from sources and resell it to consumers under varying contract arrangements. In other instances, brokers serve as an information exchange mechanism for arranging bilateral transactions between sellers and buyers.

There are sixty-five participants in the Statnett Market. The initial consequence, in a surplus energy situation, was to drive the price of energy quite low. The government became concerned over the potentially adverse effects on loans that it had guaranteed to local utilities and attempted to adjust the market by setting a minimum energy consumption of 5 GWh per year in order to be able to engage in transactions of less than five years' duration. However, the Storting, Norway's Parliament, did not accept this change.

8.4.4. Summary of Alternatives

Arguments for or against the various ways to combine or separate the major functions are beyond the scope of this chapter. The intent is to clearly show three distinctly different electricity supply industry structures. In particular we wish to call attention to the separation of the system operations and the market functions from the transmission business as illustrated in Northern Ireland. These are three distinctly different functions, and it is important to understand the relationship among them in considering the best structural design for a country or region.

There is not one and only one correct approach to arranging the major electricity supply functions. We are concerned that proposals such as were presented in the Draft Third-Party Access Directive, which was issued by the European Commission in 1992, are unduly restrictive when they require that the "transmission system operator shall refrain from buying or selling electricity" (Article 9). It is not clear that the market structure in the Netherlands complies with this requirement. The Netherlands structural model appears to be efficient and worthy of serious consideration. In a particular situation, it may not be best to separate the transmission and market functions, and it would be unwise to eliminate that option for many potentially varied situations. As always, the challenge exists for legislation to create a *direction* of change, rather than to restrict change to a specific form.

8.5. Implementation Considerations

There are a number of practical considerations that should be recognized and addressed when designing and implementing restructure of an electricity supply industry or of an individual utility. The transitional problems in achieving long-term objectives must be solved in such a way that they will not create additional problems in the future. Consequently, the following discussion presents a number of lessons that have been learned with respect to restructuring from activities of the late 1980s and early 1990s. These are described and briefly discussed in the interest of assisting in the avoidance of problems that have been experienced.

8.5.1. Market Design

There is concern that customers may be influenced by the lure of short-term spot energy prices and change suppliers. This could result in the traditional supplying utilities being left with stranded investments that must be recovered from their remaining customers, and it could also result, in the future, in insufficient commitment to new generation, in order to avoid the risk of excess or stranded investments. If utilities perceive that their future customer base is at risk, they are likely to minimize investments to either meet new demand or replace retiring generating plant.

Advocates of spot markets must understand that the existing capacity investments that currently make such markets possible were based on the perception of firm demand. They must also understand that electricity supply represents the bundling of a number of interrelated products, as indicated in the section on bulk power markets. Therefore, the purchase of any one of these products, such as energy, does not relieve the customer of the responsibility to buy or supply the rest of the products in order to maintain the prior level of quality and security of service. The conditions under which a utility must take back a customer in its service territory after that customer has gone elsewhere for supply must be clearly defined.

The market design must be adequately tested against a number of criteria before it becomes embedded in contractual cement. These tests should be conducted against a number of practical constraints including realistic long-term financial requirements and short-term operational considerations. Such testing takes time, which may not be acceptable to politically based time tables. There is, in Australia and other parts of the world, increased realization that if new markets attempt to shorten the time necessary to identify and solve problems before implementation, there is a significant risk of creating adverse commercial and technical arrangements.

8.5.2. Unbalanced Negotiations

One approach to restructure has been to permit entities within a deintegrated industry to negotiate new contractual relationships, with the government acting as arbiter. This would appear to be a reasonable and practical approach. However, there is a danger if the parties are not sufficiently balanced (that is, of equal knowledge and negotiating strength) and if the government oversight is not sufficiently knowledgeable to recognize and adjust for imbalances. The result of imbalances could be contracts that do not serve the best interests of consumers and the long-term interests of the industry. It appears that surviving generating entities from a previous structure may have an advantage over entities that were primarily concerned with the distribution and retail sales functions.

Managers and executives who may be very effective at running distribution businesses may not have sufficient knowledge of bulk power issues to engage in effective negotiations with producing entities, especially if the latter are staffed with individuals who are also knowledgeable of system operations and the operational interrelationships between demand and generation and transmission networks. There can be a tendency to not give sufficient weight to operational problems and to situations that could give a knowledgeable negotiator an advantage in establishing agreements that could be subsequently gamed.

An example is the ability of the generators in England and Wales to influence the amount of *uplift* (the cost of out-of-rate generation, transmission constraints, reserves, and ancillary services) to be paid by the retail suppliers. In the initial implementation of the England and Wales pool, this uplift was outside the prices hedged in the contracts-for-differences.

A second factor that may contribute to unbalanced negotiations is the lack of incentives for a party to negotiate in the interests of reducing costs. For instance, if distributors are permitted to pass total production costs through to captive consumers, there exists reduced incentives for the distributors to engage in negotiations with generators that will minimize costs to consumers. As always, however, the objective must not be to minimize costs in the short-term at the expense of potentially longer-term higher costs.

8.5.3. Influence of Current Supply-Demand Relationship

There is a danger of designing market structures that are unduly influenced by current imbalances in the supply-demand relationship. For instance, if there is an excess of generating capacity, it is possible to design a market that is focused mostly on energy and that does not provide a structure in which long-term capital investments may be made to finance high capital cost, new generating capacity. Financial institutions will require assurance of revenue over the investment life of a new plant. Spot markets do not provide such assurance unless they are hedged by longer-term contracts within a stable regulatory and government institution environment.

In contrast, if there is a deficiency in generating capacity, market structures may be designed to encourage development of new generating capacity. Unless appropriate market and regulatory structures are created, the danger exists of creating generating entities with substantial monopoly power. This power may be used to earn excessive profits at the expense of consumers.

The objective should be to design market structures that will promote a balance between supply and demand. The market structure should permit electric generating capacity and energy to be priced to achieve a reasonable return on investments and to protect consumers from predatory pricing practices.

8.5.4. Interrelationship Between Transmission and Generation

A major consideration that must be understood, in order to structure electricity supply industries in such a way as to minimize investment and operating costs, is the interrelationship between transmission and generation. This interrelationship is very tight: capital investments in transmission may be reduced by increasing generation production costs, and capital investments in generation may be reduced by increasing investments in transmission.

Unless this relationship is well understood, perverse incentives may be established, which result in higher costs, in generation or in transmission, than are justified. For instance, private developers of generation may prefer that a transmission network not be expanded if transmission limitations result in the need for additional generation that would be provided by competing entities. Similarly, it could be possible that the lowest total cost solution over a period of time may be installation of low-capital-cost, high-operating-cost generation, instead of a capital investment in transmission (such as in regions of the world where there is a very low density of demand). If this were the case, generation should be installed, and recovery of its costs should be allowed—to the ultimate benefit of the consumers.

It will almost always be the consumers of energy, in some cases subsidized by the taxpayers, who must pay the total costs, whether they are for generation or for transmission. However, if it is less expensive to build transmission, then private developers should not be encouraged to build generation. It is not always clear that public policy makers understand the operational interrelationships and tradeoffs between fixed and operating costs for generation and transmission and consequently may propose an industry structure or a regulatory scheme for a deintegrated electricity supply industry that may not result in the lowest total cost to consumers.

There will need to be established minimum technical standards for connection of new facilities to the transmission system. The technical standards with relation to design criteria and the construction standards for electrical facilities connected to the transmission network must be imposed contractually upon users of the transmission system. An example of such requirements are those that are included in England and Wales in the National Grid Company's grid code. However, it should also be noted that this grid code includes technical requirements that combine two major functions—system operations and transmission.

If the transmission system is provided by one entity and the system operations or control function is provided by another entity, the technical requirements for each of these functions may be provided in separate documents. As was discussed earlier, the system operations (generation scheduling and dispatch and transmission dispatch) function is not necessarily integral with provision of transmission facilities.

8.5.5. Incentive Pricing in Power Purchasing Agreements

The process for procuring generation is typically to specify an amount of capacity and solicit bids for power purchase agreements (PPAs). Therefore, the bidding process is moved one step up from what occurs when a utility builds the plant and competitively procures plant components, such as the turbine generator or the boiler. The competitive process for independent power procures, through a request for proposals (RFP), a "black box" or "generic generating plant" that provides a specified quantity of electricity, usually with specific operational characteristics. The contractual mechanism is a PPA.

In the early years of purchasing generation in this manner, not all issues were recognized or addressed in RFPs and the resulting PPAs. For instance, important requirements for integrating such new generation on a daily, hourly, and minute-to-minute basis with other sources of generation were not always included in specifications. Consequently, in some cases relatively high variable cost generation is displacing lower variable cost generation on the system. However, the trend is to recognize that energy should only be produced when its variable cost of production is low relative to other competing sources of energy production.

Adverse technical impacts, due to insufficient specification of requirements, have been minimized when the nonutility generation (NUG) has been integrated into a large existing base of utility-owned and -operated generation. However, as the quantity of generation continues to shift from what is utility-owned and operated to what is NUG-owned and -operated, it becomes more essential to negotiate contracts that will minimize the total capital and operating cost of production. The contracts must include pricing mechanisms that will send the correct signals as necessary to minimize both fixed and operating costs. It is essential that utilities have dispatch authority in terms of when generation is synchronized to the system, the level of generation that is held in reserve, the ability to control output in order to meet frequency requirements, and the ability to schedule maintenance at times when the effect on reliability and operating costs will be minimized. If utilities do not have such authority, there could be adverse effects on the consumer in terms of economy, reliability, or both.

The pitfall, which has not always been recognized, is the tendency to view a generating unit or plant in isolation from the *system* and to design contractual mechanisms without regard to the *system* and the need to integrate the operation of a single plant in a way that will ensure lowest system costs and the maintenance of reliability and system operational performance requirements. The challenge, even within vertically integrated utilities, is to communicate to plant operators that their primary purpose is to make *available* efficient generation to be loaded when called on by the system operator.

Therefore, profit incentives in power purchase agreements should primarily be tied to availability and performance targets, not to actual production. Such availability and performance targets, in turn, should be tied to when availability is of most value to the system. It makes no sense to reward production from generating units that operate at a higher variable cost than other generating units connected to the system. It also makes no sense to reward availability the same at all times. Both of these have happened with poorly designed contractual arrangements.

8.5.6. Competitive Procurement Practices

A common failing of competitive procurement mechanisms is to invite final tenders from too many bidders. Preparation of a quality proposal to develop or purchase a generating plant for a distribution utility represents a significant investment on the part of the competing entities. Procurement solicitations for large, complex projects that do not reduce the number of competitors to some reasonable number (such as five) are likely to not attract some of the most qualified bidders. The best approach is to first screen potential bidders through a prequalification process that does not require a large investment by potential bidders. Some respected developers are not attracted to the United States market because of the large number of developers which are invited to submit final proposals.

8.5.7. Metering

One of the frequent issues that needs to be addressed early in the restructure process is metering. Many times structures are proposed which are not supported by the existing metering system. For example, even though the industry in England and Wales was restructured in 1990, the metering system needed for final implementation of the designed structure was not in place until 1994. The metering necessary for implementation of full retail wheeling is still being developed.

It is possible to create elegant market structures that are based on measurements of quantities for which no metering exists. If electric facilities were not designed and built with current and voltage transformers in the required locations, a substantial amount of work may be required. The entire system of instrument transformers, transducer devices, metering recording devices, and meter data collection and processing systems may take several years to design, procure, and install.

One of the first tasks that must be carried out in a restructure, therefore, is an inventory of the existing metering. Then, as alternative structures are developed, the ability of this metering to accommodate each structure must be evaluated and both interim and permanent modifications designed. The time required for implementation of the necessary metering may, in some cases, dictate the speed of the transition.

8.5.8. Management and Regulatory Culture Change

Restructure and deintegration of the electricity supply industry requires significant changes in culture among utility managers and among regulators. The extent of the change may be great or small, depending on the extent of restructure.

Industry restructure in England and Wales required a very significant culture change for utility managers. System operators who were accustomed to scheduling and dispatching generation with the objective of minimizing the variable operating costs of all available generating units were required to perform that function based on availability offers and price quotations with an unknown relationship to actual availability and cost. The theory is that, in a competitive market, the short-run marginal price for the prices quoted should approach the short-run marginal cost or variable operating cost. However, this theory is predicated on a competitive market, with such attributes as a range of independent generating choices. When such a competitive situation does not exist, system operators may knowingly be required to issue scheduling and dispatch instructions, based on noncompetitive prices, which lead to nonoptimal costs of production.

Utility managers who have experience only within a vertically integrated structure frequently find it challenging to take on the more narrow view of one of the major functions within the industry. A manager of a distribution or transmission function, for example, may have little concern with the cost of energy that is being delivered over their facilities. A generation manager may be primarily concerned with sales and earnings and not with the issue of adequate system generating capability. In the process of developing new structures, managers need to remove themselves from the prior integrated culture and, instead, project themselves into new operating cultures.

In a similar way, regulators also must discard practices and policies that were designed for vertically integrated structures. Most of the experience in the world with price regulation of the electric supply industry resides in the United States. In England and Wales a new regulatory regime was created. It may be difficult in countries such as the United States, with an extensive regulatory infrastructure, to change policies and practices as needed to keep pace with structural changes in the electricity supply industry—especially the need to change from cost-based, or input, regulation to market-based, or output, regulation. The problem is complicated by the imperfect nature of the competitive markets in electricity and the likely continuation of some natural monopoly aspects of electricity supply. Private property rights can also make change difficult, which is why restructure should be accomplished before privatization is considered.

In countries with a well-established regulatory infrastructure that is designed for vertically integrated industries, there may be a tendency to alter regulatory processes in keeping with increased competition in a particular sector, such as generation. Public policy makers and regulators who are advocating increased coordination and optimal planning of the transmission network appear to be in conflict with the goal of achieving greater competitiveness among the entities which comprise the generation market. As noted previously, generation and transmission are closely interrelated. How does one optimally plan transmission with increased uncertainty as to the timing and location of privately developed generation?

In general, competing entities tend to want to limit information that is shared with one another. As the amount of competition in the bulk power market has increased, so has the willingness to share long-term planning information decreased. It is not clear that policy makers understand the position of entities that own generation and transmission resources and their desire to preserve commercial advantages by restricting information that would need to be shared in the development of regional planning solutions. Therefore, it is necessary for regulatory institutions, in some situations, to reduce requirements for submission of cost support information.

8.9. Summary

What is it that we expect readers to take away from this chapter? If nothing else, it is the need to go slow and consider all aspects of the industry in any effort to restructure. Additionally, we strongly encourage that the efforts associated with restructure be divorced from, and take place well in advance of, any efforts in privatization.

Restructure provides a mechanism to improve the cost effectiveness and efficiency of the electricity supply industry, whether in a whole country, a region, or a single vertically integrated utility. The ways in which the industry can be restructured are many and varied. The critical issue is to choose the final structure that will work best in a given situation. That means considering the existing structure of the industry, its current culture, the existing contractual and regulatory framework, and the goals to be accomplished.

Our hope is that readers will use this chapter as a checklist of the issues that need to be addressed. In other worlds, how does a proposed restructure address the six major functions, cross-subsidies, power supply contracts/tariffs, transmission access/pricing, competition for customers, transition issues/process, and regulation, among other others? Does the proposed industry structure provide a stable planning and operating environment for the future evolution of the industry in the interests of both consumers and owners?

Insufficient consideration of any of these aspects will only lead to further problems as any structural changes are implemented. The temptation to ignore issues or to make expedient decisions must be avoided. New structures should be extensively tested through real-world exercises in order to identify and solve problems *before* they become embedded in contractual cement. Problems and adjustments are unavoidable, but they may be reduced if the practical considerations that have been described above are addressed.

9 CONTRACT NETWORKS FOR ELECTRIC POWER TRANSMISSION[1]

William W. Hogan[2]

Everybody talks about the weather, but nobody does anything about it.[3]

9.1. Introduction

The electric utility industry has entered a new era, one characterized by competition between utility-owned and independent power, by long-term movements of power from one region to another, and by the development of short-term markets in which many buyers shop for the lowest-cost power.[4] The greater use of market forces, encouragement of new suppliers, and increasing reliance on economy power sales, plus recent precedents in power company mergers, have placed new demands on the electric power transmission systems throughout the world. Efficient use of the transmission system in a market context calls for changes in the institutions that govern transmission transactions. In the United States, "the most debated public policy issue involving the electric utility industry of the 1990s will likely be that of transmission access and the use of the bulk power system" ("Transmission," 1990, p. 12).

One unfulfilled requirement for institutional reform is a design for a

175

consistent system of (1) short-term use pricing and (2) long-term firm transmission capacity rights that can accommodate the complex problems of loop flow in the presence of line thermal limits, bus voltage tolerances, and other constraints on the transmission system.

Everybody talks about loop flow, but nobody does anything about it.[5] Most prevailing firm transmission rights are specified in terms of *contract paths* or *interface transfer capabilities* that do not address the special conditions in electric networks.[6] This chapter suggests using a *contract network* as a basic building block of a market in power transmission. A contract network and the associated rights can accommodate a system for short-term efficient pricing and long-term firm use of a transmission network.

The next sections outline the basic requirements for firm transmission rights and summarize the principal problems involved in describing the economics of electric networks, including loop flow, thermal limits, voltage tolerances, and contingency constraints. This background introduces the concept of a contract network, its associated firm transmission rights, and how long-term definition is linked to short-term efficient transmission prices determined consistent with existing network dispatch procedures. The final section outlines implementation questions raised by the prospect of designing a practical system based on a contract network model.

9.2. Firm Transmission Rights

The goal is to promote economic efficiency in the use of the electric power system. Economic dispatch of electric power plants connected through a transmission grid provides a natural starting point for discussion of efficient electricity markets. By definition, economic dispatch maximizes the benefits less the costs subject to the availability of plants and the constraints of the transmission network. The balance of costs and benefits sets prices equal to marginal costs reflecting both the direct costs of generation and the opportunity costs throughout the system. With this perspective, the work of Schweppe, Caramanis, Tabors, and Bohn (1988) develops the theory of spot pricing that respects the particular conditions of electric power transmission systems.[7] Efficient short-run prices are consistent with economic dispatch, and in principle short-run equilibrium in a competitive market would reproduce both these prices and the associated power flows.

The availability of efficient short-run prices could provide a powerful tool for guiding the use of the electric power system. The theory of spot pricing identifies the competitive price at each bus. Efficient transmission of power from one bus to another would not be priced at anything higher than the difference in the spot prices at the respective buses. Hence, this difference is the natural equilibrium definition of the price of transmission, and a matrix of spot price differences across buses provides the framework for efficient transmission pricing. Efficient prices could motivate the use of a short-run market, or, more likely, central dispatch could be used in conjunction with an efficient pricing model and a settlement system to manage the appropriate financial transfers among the participants. The theory of efficient short-run power prices provides the well-developed starting point for an efficient use of the transmission system (see Schweppe, Caramanis, Tabors, and Bohn, 1988).

Whatever the practice of short-run usage pricing, it must be integrated with a policy for long-term access and contracts for firm transmission service. From one perspective, under rather ambitious assumptions, the long-term market for power transmission could operate as a sequence of efficient short-term spot markets. The principal requirement would be for decreasing at least constant returns to scale in the transmission system. Although full reliance on short-term markets might be attractive in finessing the need for a definition of long-term transmission rights, even the narrow technical requirement of constant returns to scale is unlikely to be met in most cases.[8] Therefore, only in an ideal world would we be likely to rely solely on the optimal long-run outcome arising from a series of short-run pricing decisions.

Hence creating a competitive long-term market presents a new set of complications. And the long-run market is the key to overall efficiency. The most important requirement is to provide the right incentives for location and construction of new generating facilities and new load centers. By comparison with the costs of poor choices on these major plant investment decisions, there would likely be small operating inefficiencies from any failure to adopt a perfect short-run transmission pricing model.

In addition to assigning rights to the existing transmission system, efficient expansion of the transmission system, especially in the presence of economies of scale, presents its own set of challenges. In the absence of property or contract rights for users of the system, expansion of a centrally operated grid used by many relatively small market participants would require in principle a cost-benefit analysis that might lead to a solution that would not be replicated in a fully decentralized market. And if there are large economies of scale, efficient use of the transmission system might well produce short-run uniform prices that would not cover the cost of the expansion.[9] This subject—the optimal design and expansion of the transmission network—is an important topic that may be simplified by creation of a system of contract rights, but it is separable from the focus of the present discussion. Here we assume that there is some mechanism for deciding on the design of the system and covering the total costs (typically through "club membership" or access fees), and we address the problem of defining rights and pricing the use of the system in the short and long term.

Experience suggests that investors in long-lived, fixed facilities of the type and scale of major electric power plants will be reluctant to make commitments with no more than a promise of being allowed to participate in a short-term spot market for transmission services. Practical development of long-term deals must include some form of firm right to power transmission. Ideally there will be an associated usage pricing mechanism that reinforces the incentives for open access, economic dispatch, and efficient secondary markets for long-term firm rights.

In addition, any system for transmission rights must meet other equally important criteria. Foremost is preservation of the reliability of power system operations. Any proposal for revising the current system must recognize and respect the real complications of day-to-day management of a power network. Just as with airlines and the air traffic control system, investment and pricing rules must respect the unrestricted operational authority of the system controllers.[10] A proposal that requires a major change in current short-term system operations will face possibly insurmountable institutional barriers.

Furthermore, a reasonable transmission allocation and pricing system should be decomposable by region and company, and it must meet the test of administrative feasibility. Finally, any transmission proposal must address the Federal Electricity Regulatory Commission (FERC) concern over the existence and possible abuse of market power. Ideally a transmission protocol should be consistent with a competitive market and at least neutral with respect to the exercise of market power.[11]

No current proposal meets all the tests, and the status quo is under pressure, especially on the open access and economic efficiency tests.[12] But moving from the status quo presents a number of conceptual obstacles in defining transmission rights. What is the capacity of the system? How can we deal with loop flow and move beyond the fiction of the contract path? How can we preserve reliability and capture the benefits of efficient pricing? How can we allocate rights to the system?

9.3. Transmission Contract Path and Loop Flow

The problems created by *loop flow* are familiar to electrical engineers but often counterintuitive to others on first examination. Simply put "Because of the nature of ac [alternating current] transmission systems, energy transactions between two systems can cause flows in parallel transmission paths in other connected systems not directly involved in the transaction" (North American Electric Reliability Council, 1989, p. 31). Electricity moves according to Kirchoff's laws, essentially following the path of least resistance at the margin. As a result of these physical laws, power moves across many parallel lines in often circuitous routes.[13] These physical laws are increasingly at odds with the institutional practice for allocating transmission costs. It has been common practice to write a contract describing power flow over a specific contract path in the network. However, it may not be so easy to see that path used: "System economics often justify the use of transmission facilities among interconnecting utilities to establish "contract paths" for interchange transactions. In many instances, these facilities carry only a small fraction of the transaction, the balance of the transaction utilizing facilities in other system" (North American Electric Reliability Council, 1989, p. 36).

Hence, the *contract path* is a fiction. The actual flow of power may and often does diverge widely from the contract path. As a result, the supposed economics of the contract path may have little to do with the actual costs of the power transfer. Furthermore, these loop flows can affect third parties distant from the intended power flow, and under the current rules these third parties may and often do incur costs without compensation.

When loop flow is a small part of power economics, when informal swaps can balance out the effects over time, and when all the parties are members of the same transmission club, it is reasonable to employ the contract path fiction as a practical accommodation in crafting power contracts. These circumstances fit the past, and the contract path has been a workable fiction. But all these conditions are changing. And it is widely recognized that giving explicit attention to the economic effect of loop flow and the limitations of the contract path model presents one of the greatest challenges for designing a new power transmission regime.

The problem of loop flow is ubiquitous and can invalidate some of the most important elements of transmission agreements. For example, what is the capacity of the network? The difficulty of defining the transfer capability of the power system is closely related to the economic problems of loop flow:

"While 'transfer capabilities' between one system and another are often quoted, it is understood by those who determine them, and those who use them, that these capabilities are approximations for a specific set of conditions and not firm values that apply at all times. Therefore, a published 'transfer capability'

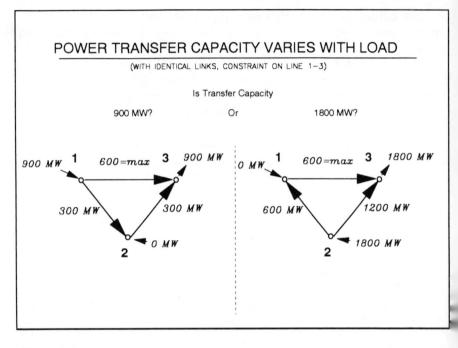

Figure 9.1.

should be regarded more as a typical or average value. The actual capability at any moment may be considerably higher or considerably lower" (North American Electric Reliability Council, 1989, p. 41).

Consider the simple example in Figure 9.1. This network consists of three buses and three lines. For the sake of illustration, assume that the lines are identical except that a thermal constraint on the line between buses 1 and 3 limits the flow on that line to 600 megawatts (MWs). Buses 1 and 2 are in the same region, perhaps the generating region, and bus 3 is the load region. What is the transfer capability from the generating region to the load region?

The two panels in Figure 9.1 depict two different load patterns that exhaust the capacity of the constrained line. As seen by a comparison of the two panels, the estimate of the transfer capability depends on the configuration of the generation. In the left panel, the total demand at bus 3 is for 900 MWs, and these 900 MWs are provided by the generator at bus 1. The flow of power follows Kirchoff's laws. Since the path 1→2→3 is twice as long as the path 1→3, it has twice the resistance. According to Kirchoff's laws, the flow is split between the two paths to balance resistance at the margin. Hence 600 MWs move along the path 1→3, while 300 MWs move along the parallel path 1→2→3. This is loop flow.[14] There is no power generated at bus 2 and none can be added there without violating the 600 MWs constraint on the line between 1 and 3. Because of the constraint, as long as we choose to generate 900 MWs at bus 1, we cannot satisfy any more demand at bus 3. In a real sense, therefore, the power transfer capability might be viewed as 900 MWs.

If demand increases at bus 3, there is no choice but to generate power at bus 2 *and* reduce the generation at bus 1; otherwise the power flow along 1→3 would exceed the maximum thermal limit. In the extreme, as shown in the right panel of Figure 9.1, if demand rises to 1800 MWs, the *only* solution is to generate all the power at bus 2 and none at bus 1. Hence, the power transfer capability might be viewed as 1800 MWs.

In a real network, the conditions in the left panel of Figure 9.1 might represent the economics and availability of generators at one time, and the right panel would apply at another. And the announced transfer capability might be somewhere in between. But "the actual capability at any moment may be considerably higher or considerably lower"—even when there has been no change in the transmission system. A contract for 600 MWs between buses 1 and 3 may have relied on a contract path along the direct connection, but the higher-demand case would have precluded this use of the system and the contract right could not be honored. Evidently changing load patterns could play havoc with transmission rights expected to extend over many years.

This small system makes the interactions apparent, but in a larger network it can be difficult to assign the loop flows. The example illustrates the problems of defining and using transmission rights along a contract path. An alternative is to define the rights in terms of a contract network.

9.4. Defining Transmission Rights in a Contract Network

If electric power flows on every parallel path, then the definitions of rights and contracts need a better approximation than the contract path. The contract network is an extension of the contract path that provides a framework for defining long-term rights while preserving short-run efficient use of the system.

9.4.1. Efficient Pricing

In the short-run, the problem of loop flow calls out for a process that will discipline use of the transmission system. In particular, the transmission protocol for a competitive market should (1) respect long-term capacity rights, (2) force short-run users to recognize opportunity costs, and (3) promote efficient trades to capture the changing economics of power loads and power generation. Those who have long-term rights should not be disadvantaged by the short-term problems of transmission congestion; those who create loop-flow congestion should pay the full cost induced by their use of the system; and those who have cheap power should be able to trade with those who have expensive power. In short, short-term transmission prices should be determined by optimal spot prices in the manner developed by Schweppe and others. Although these spot prices can vary significantly over time, the focus in this chapter is on the locational differences in prices.

The transmission prices calculated according to optimal spot-pricing theory incorporate the marginal cost of generation, the marginal cost of losses, and the opportunity cost created by congestion in the system. The first two cost groups are straightforward. Economic dispatch calls for the use of the cheapest combination of power plants needed to meet the existing load. If all the plants and loads were located at the same place, then the plants would be dispatched in order of lowest to highest marginal cost. If plants are located in different places, and power must travel over the transmission grid, then the losses of power in transmission should enter the economic dispatch calculation. But again, after adjusting for losses, economic dispatch calls for using the cheapest plants first, and optimal spot prices will be equal to marginal costs.

If the transmission grid is heavily loaded, however, bottlenecks may lead to congestion, and congestion will prevent full use of all the cheapest plants. Often referred to as "out-of-merit" or "off-cost" dispatch, the constrained use of the plants creates a frequently significant opportunity cost that can be assigned to the constraints that induce the congestion. This opportunity cost should be included in the prices.

The congestion constraints arise in two principal forms. The first and easiest to understand is the limit on the flow of power on an individual line. A change in generation or load at any bus will have some effect on the flow on the constrained line; hence, the constraint can affect the opportunity costs at each bus. It is possible to calculate the congestion cost induced by any thermal constraint and thereby estimate the effect on the optimal spot prices throughout the network.

A second major source of congestion in a power network arises from voltage magnitude constraints at buses. Even when power flows do not approach the thermal limits of the system, and the transmission lines appear to have excess capacity, voltage limits can constrain the transfer capacity and must be included in the calculation of congestion costs.

Voltage constraints inevitably require attention to both the real and reactive power loads and transfers in the alternating current (AC) transmission system.[15] Voltage can be affected by both real and reactive power loads, and the interaction between the two is critical in determining both the induced limits on real power flows and the associated spot prices.[16] In this event, spot pricing now applies to both real and reactive power, and the associated transmission prices must be determined for both types of power.

Unfortunately, voltage limitations and the associated reactive power compensation problems are prevalent. For example, the "surprising" power shortages in New England and New York in 1988 were attributed in large part to voltage problems and the "hidden but critical" role of reactive power (Zorpette, 1989, pp. 46–47). Hence it will not be enough to account for the congestion limits created by thermal limitations on transmission lines. Any new regime for transmission access and pricing must address the congestion problems created by reactive power compensation and voltage constraints. The most direct method is to establish transmission prices and contract rights in terms of both real and reactive power.

9.4.2. Capacity Rights

In other networks, long-term rights for use of the system can be defined, enforced, and traded. For example, in allocating capacity in a gas transportation system, rights can be defined to send gas through individual bottlenecks in the pipeline system. Holders of these rights could use them, in which case their gas can be identified as having flown through the bottleneck. And holders could sell the rights to others in a secondary market, in which case the buyer would have an equally well-defined asset. In a fully functioning market for the well-defined rights, prices in the short-run should reflect opportunity costs created by congestion at the bottleneck. The pipeline operator need only keep a list of the current capacity-right holders and verify that actual use of the system corresponds to the allocation of capacity. The secondary market would provide the trading opportunities and give all the participants the right incentives to pursue economically efficient exchanges.[17]

Implicitly this system for gas pipelines or other networks exploits the one-for-one definition of the capacity rights. If you sell me one unit of capacity at a bottleneck, I have one more unit of capacity. There is no complication of loop flow. But as discussed earlier in the explanation of Figure 9.1, the story is different in the case of an electric transmission network. Now there is no one-for-one trade, and the effect of different loads can be far from obvious.

However, it is not necessary to identify the trades in different uses of electric transmission capacity. As developed in the theory of spot pricing, the preferred definition of transmission and its associated pricing is from bus to bus. With transmission prices defined as the difference between the spot prices at the buses, we conceal the problems of loop flow in the calculation of the prices and marginal costs. The spot prices summarize all the information about the interactions in the network, and there is no need to define the transmission path.

The buy-sell model is an alternative interpretation of the efficient short-run pricing system that accommodates the loop-flow problem. In the buy-sell model the grid operator stands between power generators and power consumers. Ideally, the operator buys and sells power at the buses at the short-run efficient prices. One great attraction of this perspective is that there is no need to define transmission at all; users of the network never transmit power across the network; they merely sell at some nodes and buy at others. The problems of loop flow and congestion are then hidden in the internal operations of the network. All transmission is implicit. If the bus prices are short-run efficient, then the implied transmission prices are just as used here—the difference between the short-run prices at the buses.[18]

The principal difficulty with the buy-sell model as an institutional reform is in obtaining acceptance by the users of the network. If there is a strong grid with excess capacity, then the implicit transmission prices will be small and the users might be confident that the grid operator would be willing to buy and sell power at reasonable prices. But if congestion problems are large or there is uncertainty about the pricing in the buy-sell arrangement, investors might prefer a more traditional link between a plant and a customer with a well-defined transmission capability to move the power from source to destination. In this case, investors need a mechanism to protect against price changes through a definition of the capacity rights embedded in a transmission agreement.

As with the definition of transmission prices, the preferred definition of a transmission capacity right is from bus to bus, with no attention to the paths by which the power flows. Hence, in Figure 9.1 we might define the generator as bus 1 as having a "right" to send 900 MWs to bus 3. But suppose conditions change, and it is cheaper or necessary to generate 1800 MWs at bus 2, as shown in the right panel in Figure 9.1. How can we organize events such that the capacity-right holder located at bus 1 sells the right to transfer 900 MWs and the generator at bus 2 buys the right to transfer 1800 MWs?

One approach might be to allow for bilateral trades of capacity rights among users of the network. For the simple three-bus case, it is an easy matter to see the interactions among users of the system. But as the network gets more complicated than that shown in Figure 9.1, there will be many intervening buses and lines. With rights for transmission capacity defined between bus pairs, the complexity of the interactions and required trades could eliminate all but a few one-for-one trades, and the problem of loop flow would return to the forefront. Obviously, the central grid operator, someone with an overview of the system, must be involved. But that involvement must be more than just keeping a list of the capacity-right holders.

In principle, the central operator would know or be able to calculate that 1 MW from 1 to 3 displaces 2 MWs from 2 to 3. Hence, one approach would be for all trades of capacity rights to be made through the central operator with the requirement that before any power moves the appropriate capacity must be obtained from the capacity-right holder. This form of a buy-sell model with the grid would allow efficient trades and give users the necessary information about opportunity costs. But it would place a substantial burden on the central operator, especially if there were many trades. Regrettably, it is normal for economic and load conditions to change frequently, so that there would be a requirement for many capacity trades.

An alternative would be to merge the short-term pricing and an implicit secondary market in capacity rights as part of a contract network. Here a one-step pricing method is available that will allow for implicit trades without imposing new demands on the operator of the transmission grid. In particular, given a designated "swing" bus, recall that the spot price at any bus is a combination of the system marginal cost of generation at the swing bus, the effect on losses, and the effect on congestion. Hence, for bus i,

$$\text{Bus price}_i = \text{Generation} + \text{Losses}_i + \text{Congestion}_i \qquad (9.1)$$

The transmission price between any two buses can be defined as the difference in the spot prices. Consequently, the transmission price between bus i and bus j is

Transmission price$_{ij}$ = (Losses$_j$ − Losses$_i$) + (Congestion$_j$ − Congestion$_i$),
$$\tag{9.2}$$

or

$$T_{ij} = T_{Lij} + T_{Cij}. \tag{9.3}$$

The first term, T_{Lij}, is just the incremental effect of transmission from bus i to bus j on the system losses and can be thought of as the operating cost of the transmission service. It would be normal to expect both long-term and short-term users to pay this cost.

The second term, T_{Cij}, is the incremental effect of transmission from bus i to bus j on the system congestion costs. This charge has an interpretation as the rent on the transmission capacity used by the power transmission from bus i to bus j. Since the congestion charge measures the opportunity cost, this is the maximum that others in the system would be willing to pay to purchase the capacity right. If a holder of a capacity right actually uses the transmission system, then at the margin efficiency would require that the right holder recognize that trades are available that would pay T_{Cij} for the capacity right between bus i and bus j. Similarly, if others use the transmission system and through loop flow prevent the capacity-right holder from transmitting power, then the T_{Cij} charge from the optimal transmission price is the minimum that the right holder should accept to "rent" the capacity to the actual users.

These interpretations of the congestion charge suggest a definition of the capacity right that would be consistent with successful operation of a secondary market. For any period, suppose that the loads and transmission flows follow economic dispatch and the corresponding spot prices are available consistent with economic use of the system. Then all users of the transmission system are charged the transmission prices T_{ij} according to their usage. But in addition, the owners of capacity rights receive a "rental" payment from the grid equal to T_{Cij}, the congestion charge, applied to their full capacity right. Hence, everyone pays both the loss and the congestion charge, so everyone faces incentives for efficient short-run use of the system. And the capacity-right holder is compensated if loop flow or load conditions prevent full exercise of the capacity right.

If the right holder uses the full capacity right, then the congestion charge paid is just balanced by the rental payment, and the net cost of transmission is just the losses charge. This is true no matter what the load conditions or no matter how large the total transmission charge.[19]

But just as important, in the presence of optimal spot prices, whenever the right holder is precluded from using the full capacity, the compensation received is just the amount needed to make the right holder indifferent between delivering the power or receiving the compensation.[20] In the latter case the right holder can honor any long-term delivery commitments by using the rental payment or congestion fee to purchase expensive power at the point of destination. The congestion charges could arise from thermal limits on lines, voltage constraints at buses, or a mixture of both. Since they derive from optimal spot prices, they are consistent with any efficient pricing mechanism. In other words, the calculation of the optimal spot prices for the contract network in the short-run produces the same result as the ideal secondary market but without the requirement for explicit capacity trades.

At the margin, both the short-run user and the capacity-right holder would face the same incentives, but the capacity-right holder would also receive a rental payment that guarantees the economic viability of long-term power sale requirements. Hence, both the transmission capacity rights and the transmission prices can be defined between pairs of buses.[21] And there is no need to identify any transmission contract paths. The problems of loop flow are again handled in the calculation of the optimal spot prices for the contract network.

9.4.3. Calculating Prices

As shown by Schweppe, Caramanis, Tabors, and Bohn (1988), optimal spot prices could be obtained as a byproduct of economic dispatch and real-time spot pricing could be used to control the network. Such real-time spot pricing would be a major innovation that would simplify much of the effort to achieve theoretical conformance with the principles of efficient use of the system. However, system operators already have well-developed methods for controlling the network, and they currently seek an economic dispatch within the limitations of a variety of explicit and implicit constraints. Given the importance of close control of the network to maintain reliability, there would be natural and legitimate objections to imposing dramatic changes in the dispatch methods in order to simplify estimation of transmission prices.

There is merit in improving the dispatch process, if this can be done. But the dispatch process is both complicated and critical in meeting the essential reliability standards of the transmission network. When trying to introduce the use of spot power prices for calculating efficient transmission prices, therefore, a less ambitious goal would be in order. One approach is to accept as given the results of the dispatch process and then allow for periodic or even ex post estimation of consistent spot prices, all the while accepting the actual system dispatch as an optimal balance of the underlying economics and constraints but leaving the dispatch process undisturbed.[22] The dispatchers would be presumed to be solving a difficult problem, and the prices would be calculated to communicate the right incentives to the transmission customers.

Even when there is no central computer calculating the optimal prices, with these prices in turn guiding dispatch, there usually is enough information available to indicate the implied constraints on the network. This description of the results of the dispatch process depends on information that is familiar to dispatchers and can be specified without extensive calculation. And with this limited amount of information, there is an easy method for estimating prices under the assumption that the actual dispatch is the result of an optimization with respect to a set of constraints.

The additional information includes an identification of the binding constraints in the transmission system and the variable costs of operation of the marginal generating plants at critical buses. Typically the system operator has a good deal of experience with the critical constraints in the transmission network and ready knowledge of the short-run variable costs of operation of plants. We need know only the identity of the constraints and do not require ex ante estimates of the limits on the flows or voltages.

For the costs at critical generating buses, even if they are not reported regularly, we can reasonably estimate these costs, which should be dominated by energy costs. At each bus we will know the plants, or more precisely the units, that are running and those that are available but idle. Then the running cost of the most expensive plant in use but running at its upper limit provides a lower bound on the spot price at that bus. Apparently if the true spot price (including the import or export of power) were lower, the plant would not be dispatched. Likewise, the running cost of the least expensive plant not running at that bus provides an upper bound on the spot price at that bus. Apparently if the spot price were higher, the plant would be running. And in the likely event that the marginal plants are partially loaded, the incremental running cost provides both lower and upper bounds and thereby determines the spot price at that bus.

For locations where there is direct customer load, additional information may be available to obtain better bounds on the spot prices. For example, interruptible supply contracts might provide lower bounds if the supply had been interrupted and upper bounds otherwise. Or there may be data available on outage costs to provide an upper bound on the spot price at any location. Obviously, the better the information and the tighter the bounds, the better the estimation of the spot price.[23]

This type of variable cost information may even be reported as part of a billing and payments scheme for generators. For instance, the commonly used split-savings systems in power pools depend on estimates of the marginal cost of all plants, both those run and those left idle. Hence, electric utilities have demonstrated the ability to provide acceptable estimates of the key information. In any event, collection or reasonable estimation of these data should be a modest additional burden when reporting the net loads at each bus. And given this additional information, transmission prices can be estimated ex post. This ex post method is particularly attractive as a transition approach for developing a new transmission regime. The ex post method allows the current dispatch operations to remain in place and calculates prices consistent with the actual usage by applying the marginal tests of economic dispatch.

Hence, the contract network can operate within the existing network control system, requiring a reform only in pricing when combined with a settlement process to redistribute the transmission congestion payments. The redistributed congestion charge makes the capacity-right holder indifferent between transmitting the power and keeping the rental payment.

9.5. Example Contract Network

The simple three-bus and three-line network from Figure 9.1 will serve to illustrate the pricing and payments in the contract network (see other examples in Hogan, 1990). Assume that this network has been accepted as a reasonable approximation of the major regions and connections for the underlying system. Furthermore, assume for simplicity that the load is always met, sometimes by operating expensive generation facilities, and that operators provide an economic dispatch subject to the constraints on the system. Then after identifying the net loads at each bus and the binding constraints for the applicable contingency, either thermal constraints on lines or voltage constraints on buses, ex post prices can be calculated as the optimal spot prices.[24]

For purposes of the illustration, we concentrate on the case of a thermal limit on the line from bus 1 to bus 3. Figure 9.2 repeats the network but adds the bus prices. Here the left panel describes both the initial load and the allocation of capacity rights. Hence, a generator at bus 1 is assumed to have the right to transfer 900 MWs of power to bus 3. The load at bus 3 is just satisfied by this transfer. Furthermore, the cost of power at bus 2 is assumed to be too high to be economic, so there is no need to transfer power from bus 2 to bus 3.

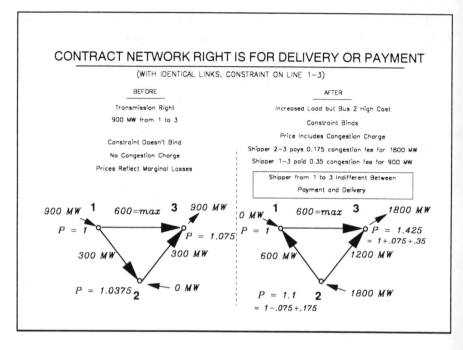

Figure 9.2.

Under these conditions, the 600 MWs constraint on line 1 to 3 is not binding. As shown in the left panel of Figure 9.2, the prices consist only of the charge for marginal losses. There is no congestion charge. Note that for the quadratic direct current (DC) load approximation, marginal losses are always linear in flow. The flow on path 1 to 2 to 3 yields marginal losses of 0.075, and along path 1 to 3 the losses are also 0.075. The path 1 to 2 to 3 is twice as long, but the path 1 to 3 has twice the flow. Hence, the marginal losses equalize along each path.

Normalizing for the price at bus 1 (the swing bus), the ex post optimal spot prices are 1.0375 at bus 2 and 1.075 at bus 3.[25] And the transmission charges are 0.075 (i.e., 1.075 − 1.0) for power transmitted from 1 to 3, and 0.0375 (i.e., 1.075 − 1.0375) for power transmitted from 2 to 3. The only power flowing is the 900 MWs moving from 1 to 3, so the total transmission payment is 900*0.075 = 67.5. The capacity-right holder pays this full transmission cost. And since there is no congestion charge, the rental payment for the capacity right is exactly zero.

If the system operates only in the mode of the left panel of figure 9.2, then the capacity-right holder always enjoys full access to the system, pays only the operating cost to cover marginal losses, and there is no congestion rental payment. But suppose now that after the assignment of capacity rights there is a change in system economics and load conditions. Suppose now that the demand for power at bus 3 increases to 1800 MWs as shown in the right panel of figure 9.2. Here the constraint on line 1 to 3 is known to be binding. Furthermore, suppose the dispatch required use of relatively expensive generation at bus 2 and the system operators place a lower bound on the price at bus 2 of 1.1 relative to the price at the swing bus. We assume that the operators know or can estimate these three sets of information: the net loads at the three buses (0 MWs, 1800 MWs, −1,800 MWs); the identity of the binding constraint (we know the constraint binds on the line from 1 to 3 but do not need to specify the exact flow on the line); and a bound on the price at one of the two nodes (price at bus 2 must be at least 1.1, the cost of the most expensive plant actually generating power at this bus).

Then the ex post price calculation determines the optimal spot prices as 1.0, 1.1, and 1.425. Furthermore, the same calculation separates the prices into the contribution of generation (1.0, by definition), marginal losses (0, −0.075, and 0.075), and congestion charges (0, 0.175, 0.35).

Why does the pricing model yield these new estimates? The marginal losses of (0, −0.075, and 0.075) can be verified by noting that the flow along 1 to 3 is the same as before, so the marginal losses along this line must not have changed. The reduction at bus 2 to the net of 0.925 is obviously the value such that the marginal losses from bus 2 to bus 3 equalize on either path with the new flows. In other words, for this dispatch to be optimal in the absence of the constraint, the cost of power at bus 2 would have to be nor more than 0.925.

The relative price at the swing is 1.0 by definition. And the price 1.1 at bus 2 is determined by assuming that an expensive plant is running "out-of-merit" at a known relative price. The constraint induces a congestion charge at bus 2 of 0.175, (1.1 − 0.925 = 0.175). The remaining price at bus 3 is calculated in the pricing model. However, it can be verified by tracing the changes and costs savings if load at bus 3 reduces by 1 MW.

With a 1 MW reduction at bus 3, we would need 1 MW less at bus 2. In addition, the dispatcher would be able to substitute 1 MW at bus 1 for an additional 1 MW at bus 2 and still not violate the constraint of 600 MWs on the line 1 to 3. After this redispatch, the flows would be 1 to 3:600, 1 to 2:−599, 2 to 3:1199. The resulting net loads would be bus 1 at 1 MW; bus 2 at 1798 MWs; and bus 3 at −1799 MWs. We would save 2.2 at bus 2, pay 1.0 more at bus 1. In addition, we would save 0.075 of losses on 1 to 2. Since the flow on line 2 to 3 is twice that on line 1 to 2, the marginal losses are also twice, yielding a loss saving of 0.15 on 2 to 3. The total saving is $2.2 - 1.0 + 0.075 + 0.15 = 1.425$. And this 1.425 is just the price at bus 3. Hence the price represents the full opportunity cost with the optimal dispatch in the presence of the constraints.

With this calculation, which generalizes to a more complicated network, we determine the congestion rentals as the differences between the generation and loss-only prices (1.0, 0.925, and 1.075) and the full-bus prices (1.0, 1.1 and 1.425). From these spot prices we can derive the optimal transmission charges as

$$T_{13} = (1.425 - 1.0) = T_{L13} + T_{C13} = (1.075 - 1.0) + (0.35 - 0), \quad (9.4)$$

or

$$T_{13} = 0.425 = T_{L13} + T_{C13} = 0.075 + 0.35, \quad\quad\quad\quad (9.5)$$

and

$$T_{23} = (1.425 - 1.1) = T_{L23} + T_{C23} = (1.075 - 0.925) + (0.35 - 0.175), \quad (9.6)$$

or

$$T_{23} = 0.325 = T_{L23} + T_{C23} = 0.15 + 0.175. \quad\quad\quad\quad (9.7)$$

However, the capacity-right holder is unable to use the 900 MWs transfer right. As examination of the congestion charges shows, the rental payment is just the amount needed to honor the 900 MWs delivery requirement. The actual user pays 0.325 for each of 1800 MWs shipped from bus 2 to bus 3. The congestion charge included in this transmission price is 0.175 for a total rental payment of $0.175*1800 = 315.0$. And this rental payment is just equal to the payment that the grid makes to the capacity-right holder of the 0.35 congestion rental from 2 to 3 for the 900 MWs, or $0.35*900 = 315.0$.

Furthermore, if the market is in equilibrium at the economic dispatch, which is an explicit assumption of the optimal spot-price calculation, then the cost of delivered power at bus 3 is 1.425. Presumably, the generator at bus 1 has an obligation to meet 900 MWs of the 1800 MWs load at bus 3. At the current prices, the generator is indifferent between actually generating at bus 1, delivering the power and paying the transmission price of 0.425, or purchasing the power at bus 3 at a price of 1.425. The capacity-right holder still retains the rental of 315, which makes the effective costs of transmission equal to the cost of losses only. Viewed another way, the right holder from 1 to 3 is indifferent to using the system and paying the marginal cost of losses, for an effective delivered price of 1.075; or purchasing the power at bus 3, which after accounting for the rental payment of 0.35 also yields an effective price of 1.075.

This three-bus example illustrates the key principles of the contract network. From here it is a straightforward matter to extend the application to a larger, more realistic network that includes real and reactive prices across many locations with contingency constraints and other important operational feature (for details and examples see Hogan, 1992a).

9.6. Implementation Questions

Any implementation of a contract network would raise a number of questions. As with other applications of marginal-cost pricing, the general framework would accommodate many alternative formulations based on the same underlying principles. The innovations would require institutional changes in either the national grids of other countries or the Balkanized grid in the United States. Rate design and capacity expansion decisions would change in the presence of a system of capacity rights.[26] And the description of the contract network goal leaves open transition questions such as the treatment of existing rights.

The contract network definition of capacity rights requires some allocation of those rights. In principle, any feasible allocation would be compatible with the ex post pricing system and reallocation of congestion rents. Furthermore, the contract network framework appears neutral with respect to the choice between voluntary systems preferred by transmission owners (Rueger, 1990s, pp. 36–37) and the default mandatory access sought by new entrants (Penn, 1990, pp. 26–30). In practice, the assignment of initial rights will be a complicated task. The existing transmission system has been paid for by existing users, and there will be many assertions of grandfathered rights to the current system.

A full investigation of the alternative methods of allocation, recognizing both historical obligations and the pressure for economic efficiency, is a separate topic that goes beyond the present discussion. However, there is one natural alternative that follows from the same observations that support the development of the contract network. Given the use of the contract network to integrate the calculations of transmission capacity usage and the effect of loop-flow, it is natural to suggest the use of a similar mechanism for allocating capacity.

Suppose that all possible pairs of generation and consumption options are included implicitly or explicitly; then a set of bids for transmission capacity from bus to bus could be identified with an associated maximum price that could be paid for each right. The generic model used in calculating the short-term prices and congestion rents could be adapted with a new objective function—namely, maximizing the value of the selected bids, to determine a combination of bids that would (1) be feasible, (2) recognize the interactions of loop flow, and (3) provide the most efficient allocation of transmission rights.

Use of a contract network, with or without the formal bidding model for allocation of initial capacity, leaves open important questions in the design of a complete transmission protocol. For example, refunding the congestion rents leaves the grid with only the rents on losses, which are not likely to cover the fixed costs of the existing transmission system. This will create the need for the design of a system of access fees, operating in parallel with the short-term pricing of losses and congestion in the contract network.

For similar reasons it would be unusual for the natural incentives associated with the contract network to align the required decisions for optimal expansion of the transmission grid. In the presence of economies of scale, the expectation would be that the overall benefits of expanding the grid would be large but might not be sufficient to justify the expansion for recipients of the new capacity rights based solely on the revenues that would be collected through the congestion rents. Although the contract network would provide property rights to eliminate the loop-flow-related problems of the commons, free riders might wait for others to expand the grid in the hope of relying on the subsequent unconstrained spot market. In this case, a cost-benefit analysis and public oversight through regulation would typically be required to support system expansion decisions.

The related problems of cost-based rate regulation for transmission may be simplified by the likely excess of average over marginal costs. This might provide the opportunity to accommodate a contract network pricing system that reflects short-run opportunity costs and a long-term access charge that nets out any aggregate opportunity rents. In this circumstance, there would be cost-based rates with the proper incentives for use of the transmission system.

The idealized contract network with the market equilibrium assumptions implies that we can calculate the efficient prices. However, it is easy to imagine the existence of pockets of market power where participants in the system might be able to manipulate price and cost data to their advantage. It remains as a subject for future research to determine the robustness of the contract network operations in the face of varying degrees of market power.[27] However, as mentioned above, the difficulties with a contract network should be no more severe than for the common split-savings systems that have been an accepted practice for many years. Furthermore, by providing a viable long-term transmission right, a contract network could help mitigate market power.

Similarly, the contract-network framework builds on the theory of spot-pricing from Schweppe, Caramanis, Tabors, and Bohn (1988), but the specific requirement is only for calculation of the *differences* in the efficient prices at the buses. Hence, at least within the network if not at the boundaries, actual power sales might be based on any of a number of contract or regulatory provisions that deviate from marginal cost pricing. Assuming economic dispatch, it is an open question as to how much these power pricing rules can differ from pure spot pricing without subverting the incentives for efficient transmission.

Contrary to conventional practice—as, for instance, in the gas pipeline capacity allocation model—reliability or priority categories appear to be moot in a contract network as defined here. Ordinarily priority categories address the problem of specific performance. If the capacity is unavailable, the lower priority capacity-right holder is curtailed and enters the secondary market to purchase alternative supplies or capacity rights. But as the mechanism for accommodating the problems of loop flow, the contract network rights integrate specific performance and the putative result of trades in a secondary market. Any capacity-right holder keeps a rental payment if curtailment is necessary, and the rental payment is always determined by the value of the constrained capacity in the short-term market. Hence the contract network appears to be an alternative to the use of priority categories or reliability blocks.

The basic contract-network model is designed for a single capacity profile with coincident requirements for transmission capacity. It may be that a single allocation of capacity is sufficient for a transmission system, or there may be a preference for different allocations according to expectations about different coincident peaks. Issues such as the appropriate period of coverage remain as empirical questions that would be addressed best in the context of particular transmission grids.

9.7. Conclusion

A contract network provides an internally consistent framework for defining long-term capacity rights to a complicated electric transmission network consistent with the principles of a competitive market. By design, a contract network would maintain short-run efficiency through optimal spot-price calculation of transmission prices. Through the payment of congestion rentals, the contract network makes the capacity-right holder indifferent between delivery of the power or receipt of payments in a settlement system. The contract network respects the special conditions induced by Kirchoff's laws and the prevalence of loop flow, thermal limits, voltage constraints, and contingencies. Furthermore, a contract network approach can be adopted without necessarily disturbing existing methods for achieving an economic power dispatch subject to these constraints.

Notes

1. This is an abridged version of Hogan (1992a).

2. Thornton Bradshaw Professor of Public Policy and Management, Kennedy School of Government, Harvard University, and director, Putnam, Hayes & Bartlett, Inc., Cambridge, MA. I have benefitted from repeated conversations on transmission pricing with members of the Harvard Utility Forum, colleagues and clients at Putnam, Hayes and Bartlett, and many others including Robert Arnold, Homer Brown, Douglas Bohi, Roger Bohn, Bernard Cherry, Charles Cicchetti, Ron Clark, Gordon Corey, James Cunningham, Charles Davies, Ken Fleming, Richard Flynn, Mark Friese, James Groelinger, George Gross, Kenneth Haase, George Hall, Steve Henderson, Steve Herod, Eric Hirst, Terry Howson, Robert Irwin, Joseph Keppinger, Henry Lee, William Lindsay, Cathy Mannion, David Marshall, John Macadam, James Malinowski, John Meyer, Thomas Milburn, Ray Orson, Howard Pifer, Douglas Powell, Martin Rosevear, Bart Smith, Charles Stalon, Irwin Stelzer, Donald Stock, Hodson Thornber, and Max Wilkinson. The idea of using contract networks for defining long-term rights grew out of intensive discussions with Sarah Johnson, Thomas Parkinson, Larry Ruff, and Michael Schnitzer. Support has come in part from the Harvard Utility Forum. The author is a consultant on electric transmission issues for Duquesne Light Company, the British National Grid Company, and Electricorp of New Zealand. The views presented in this chapter are not necessarily attributable to any of those mentioned, and the remaining errors are solely the responsibility of the author.

3. Often attributed to Groucho Marx, but earlier from Charles Dudley Warner, Editorial, *Hartford Courant*, August 24, 1897.

4. Joskow (1989) provides an overview of the recent trends of competitive forces in the United States. For a comparison of U.K. and U.S. trends and the impact on transmission, see Wilkinson (1989).

5. For an extended discussion of the problem of loop flow and other special conditions in electric markets, see Federal Energy Regulatory Commission (1989). The National Regulatory Research Institute report (Kelly, Henderson, and Nagler, 1987) provides an excellent introduction and overview of the principal technological, economic, and institutional factors relevant to transmission networks.

6. One partial exception is the Texas region, ERCOT, which uses network flow analysis to evaluate wheeling and transmission under certain simplifying assumptions to calculate impacts throughout a network. The AC system is electrically isolated from the rest of the United States.

7. The bible for the development and summary of the theory of spot market pricing is Schweppe, Caramanis, Tabors, and Bohn (1988). For a summary of the theory and related analyses, see Hogan, (1992).

8. The problems created by economies of scale are addressed in the context of transmission pricing in the New Zealand by Read (1988) and Read and Sell (1988a, 1988b).

9. With a single price for use of the system, this is the familiar bridge-toll-setting problem where the total benefits as measured by consumer surplus outweigh the costs of the bridge, but the optimal short-run price would not produce enough revenue to cover those same costs.

10. Eric Hirst suggested the analogy.

11. In their innovative investigation of the even more ambitious step of deregulating *existing* power plants, Schmalensee and Golub (1984, p. 21, emphasis in the original) found that "the estimates of effective concentration are *extremely* sensitive to the transmission capacity assumption."

12. For a summary of alternative decisions and proposals, see Federal Energy Regulatory Commission (1989). This chapter was prepared before passage of the Energy Policy Act of 1992, which as enacted contains the authority to mandate transmission access for wholesale transactions. Since the law does not specify how to implement transmission access and pricing, the new legislative authorities provide further motivation for the analysis provided here.

13. The FERC (1989) report contains an excellent discussion of the problems of loop flow.

14. Here we are ignoring losses and use the conventional DC load approximation for purpose of the illustration. Since the lines are identical, path 1 to 2 to 3 has twice the resistance of path 1 to 3, which makes it easy to verify the power flows.

15. See Kelly, Henderson, and Nagler (1987). Chapters 2 and 3 provide a usable description of the role and impact of reactive power in transmission networks.

16. It is well known that real power flows are largely determined by the difference in voltage angles across lines and the flow of reactive power is determined by the differences in voltage magnitudes across lines. This often leads to a "decoupled" analysis of power flows, and this allows the development of spot prices for real power without considering reactive power loads. Furthermore, computational procedures for solving the AC load-flow equations often exploit a similar procedure. However, the decoupling on flows via the differences in angles and magnitudes does not apply for the angles or the voltage magnitudes themselves, and calculation of induced constraints and spot prices depends importantly on recognizing the interactions among real power, reactive power, and voltage magnitude. See Hogan (1992a) for further details and references.

17. For a further discussion of a similar concurrent auction mechanism and the connection to competitive markets, see Hogan (1992b). The model can be thought of as a bundle of straws from well-head to burner tip. In practice, exchanges and commingling of gas make this only an idealized approximation. But it is a workable approximation that allows for the definition of rights, pricing, and contracts.

18. This buy-sell model is close to the design adopted for the British national grid, but the British pricing system for buses does not recognize congestion constraints or the marginal costs of out-of-merit dispatch. Hence the implicit transmission pricing does not yet provide efficient incentives for location of new generation plants. This transmission pricing scheme is of course subject to further revision in the still incomplete transition in the privatization of the British electric system.

19. For simplicity, the focus here is on the congestion charge, which is the greatest source of price variability. However, as proposed by Larry Ruff, the ideas can be extended to treatment of increased or decreased charges for marginal losses.

20. This use of a contract to substitute payment for specific performance has a close analogies in the power supply contracts in the evolving British market. Sales are to and from the grid, but side payments between producers and consumers provide a near perfect hedge

that make the economics look like a direct sale from the individual producer to the customer. With the introduction of transmission constraints and pricing, the capacity contract provides the analogous financial hedge against price changes. Of course, the capacity right is strictly a financial instrument and confers no physical control over the network. To emphasize that there is no attempt to interfere with operations, alternative terminology might be "nonexclusive rights to compensation" as suggested in New Zealand by Grant Read, or "congestions contracts" as described in the United Kingdom by Putnam, Hayes and Bartlett, Ltd.

21. Other equivalent formulations could be developed. For instance, each right between any two buses could be decomposed into a pair of rights relative to a reference bus. In this formulation, all "transmission" would be to or from the reference bus. This would permit such simplifications as assigning generators rights to the reference bus and customers rights from the reference bus.

22. Roger Bohn points out that ex post pricing of transmission services would be similar to the familiar ex post pricing through fuel adjustment clauses in the United States.

23. Schweppe, Caramanis, Tabors, and Bohn (1988, p. 97) argue for the robustness of spot pricing even when the perfect optimal solution is not available: "The fact that the true (prices) may not be calculated does not destroy the value of implementing a spot price based energy marketplace. The actual value calculated will be much closer to the true values than the present-day flat or time-of-use rates, etc. The goal of implementing the spot price based energy marketplace is to improve the coupling between the utility and its customers, not to achieve theoretical optimality."

24. See Hogan (1992a) for a description of the method of ex post calculation of the prices given loads and the location of any binding constraints.

25. The specific prices depend on the details of the network. However, the base case is not important for developing the example (1990).

26. For example, a contract network could meet many of the pricing criteria set out by Commissioners Ashley C. Brown and Terrence L. Barnich (1991) in their search for an alternative to ratebase treatment of transmission assets.

27. See, for example, the discussion in Einhorn, (1990). Abstracting from the problems of networks and loop flow, Einhorn develops nonuniform price-cap models that provide incentives for profit-maximizing utilities to provide the welfare-maximizing transmission capacity, even in the presence of market power. However, as Einhorn notes, the method depends in part on the ease of analysis of the wheeling customers' profits and may be "difficult if many wheeling customers appear," as is the case in the typical network.

References

Brown, Ashley C., and Terrence L. Barnich. 1991. "Transmission and Ratebase: A Match Not Made in Heaven." *Public Utilities Fortnightly*, June 1, pp. 12–16.

Einhorn, M. 1990. "Electricity Wheeling and Incentive Regulation." *Journal of Regulatory Economics*, 2: 173–189.

Federal Energy Regulatory Commission. 1989. *The Transmission Task Force's Report to the Commission: Electricity Transmission—Realities, Theory, and Policy Alternatives.* Washington, DC: FERC.

Hogan, W. 1990. "Contract Networks for Electric Power Transmission: Technical Reference." Discussion Paper E-90-17. Energy and Environmental Policy Center, Harvard University, September.

———. 1992a. "Contract Networks for Electric Power Transmission." *Journal of Regulatory Economics*, 4 (September): 211–242.

———. 1992b. "An Efficient Concurrent Auction Model for Firm Natural Gas Transportation Capacity." *INFOR (Information Systems and Operational Research)*, 30(3) (August): 240–255.

Joskow, P.L. 1989. "Regulatory Failure, Regulatory Reform, and Structural Change in the Electric Power Industry." *Brookings Papers on Economic Activity: Microeconomics* (pp. 125–208). Washington, DC: Brookings Institution.

Kelly, K., J.S. Henderson, and P.A. Nagler. 1987. *Some Economic Principles for Pricing Wheeled Power.* NRRI-87-7. Columbus, OH: National Regulatory Research Institute.

North American Electric Reliability Council. 1989. *1989 Reliability Assessment.* Princeton, NJ: NERC, September.

Penn, D. 1990. "Movement Towards an Efficient and Equitable Transmission Policy." *Public Utilities Fortnightly*, July 19, pp. 26–30.

Read, E.G. 1988. "Pricing of Transmission Services: Long-Run Aspects." Report to Trans Power, Canterbury University, New Zealand, October.

Read, E.G., and D.P.M. Sell. 1988a. "A Framework for Transmission Pricing." Report to Trans Power, Arthur Young, New Zealand, December.

Read, E.G., and D.P.M. Sell. 1988b. "Pricing and Operation of Transmission Services: Short-Run Aspects." Report to Trans Power, Canterbury University and Arthur Young. New Zealand, October.

Rueger, G. 1990. "The FERC's Transmission Task Force Report: Where Do We Go from Here?" *Public Utilities Fortnightly*, February 1, pp. 36–37.

Schmalensee, R., and B. Golub. 1984. "Estimating Effective Concentration in Deregulated Wholesale Electricity Markets." *Rand Journal of Economics*, 15(1) (Spring): 21.

Schweppe, F.C., M.C. Caramanis, R.D. Tabors, and R.E. Bohn. 1988. *Spot Pricing of Electricity.* Norwell, MA: Kluwer.

"Transmission: A Continuing Controversy." 1990. *Public Utilities Fortnightly*, July 19, p. 12.

Wilkinson, Mas. 1989. "Power Monopolies and the Challenge of the Market: American Theory and British Practice." Discussion Paper E-89-12, Energy and Environmental Policy Center, Kennedy School of Government, Harvard University.

Zorpette, G. 1989. "Moving Power Through the Northeast Corridor." *IEEE Spectrum* (August): 46–47.

10 A CONCEPTUAL REGULATORY FRAMEWORK OF TRANSMISSION ACCESS IN MULTIUTILITY ELECTRIC POWER SYSTEMS

Ignacio J. Pérez-Arriaga

10.1. Introduction

There is an ample consensus about the existence of significant economies of scale in the *coordination* (which does not necessarily imply centralization) of the operation and capacity expansion planning of electric power systems. It is believed that, in order to capture all the coordination-related economies of scale, very large systems have to be considered. A figure of several tens of gigawatts (GW) of installed generating capacity has been mentioned as the minimum size that is needed to exhaust these economies (see Joskow and Schmalensee, 1983). Recent studies within the context of the European Union (EU) internal market for electricity seem to indicate that much larger systems may still benefit from coordination practices. These systems will typically encompass several utilities and, in most cases, they may also involve a number of different countries.

The introduction of market forces to improve the economic efficiency of the electric power industry is a regulatory approach that has been decidedly used in some countries, such as Chile (Bernstein, 1988) and recently the United Kingdom, and it is presently widely considered in many parts of the world. Promoting *competition* has been frequently understood as a twofold scheme: (1) allowing the generating units the possibility of selling their product anywhere and (2) providing the buyers of electricity (large consumers and distribution utilities) with access to the lowest electricity prices—what has been perceived as equivalent to the possibility of purchasing electric energy from any willing supplier.

Realizing economic efficiency through coordination and competition appears to be critically dependent on the absence of impediments to the trade between willing participants and, in particular, to the possibility of *access* to the interconnected transmission network. However, there is a universal lack of consensus on the most adequate regulation of transmission access in every case where this issue has been raised, whether wheeling in the United States of America or third-party access in the EU internal electricity market. Pricing of transmission services is a particularly involved and related issue, which is also far from being settled. The most frequently expressed concerns about open transmission access regulation are the following (see, for instance, Commission des Communautés, Européennes, 1991; Edison Electric Institute, 1986; Current Operational Problems Working Group, 1991; Adamson et al., 1991):

- *Security* of power supply must be preserved at least at the existing (and in general satisfactory) levels. It has been argued that extending the scope of the systems to be coordinated and allowing a multiplicity of transactions between the numerous parties in highly deregulated power systems will have a negative effect on system security because of the ensuing difficulties in operation and planning.
- Preservation of the *autonomy* of operation and investment decisions is frequently an essential requirement of some participants in an electricity marketplace. This is likely the case when the participants belong to different countries.
- Because limited transmission access is associated with some level of *market power*, some participants presently enjoy some advantages that will be lost if regulation access is liberalized. Therefore resistance to these changes in regulation is to be expected. A particularly serious situation is what has been called the stranded investment case, whereby a generating group that is shielded from competition because of limited transmission access may become underutilized in a more competitive setting, resulting in reduced revenues that may be very short of recovery of the total (fixed plus variable) costs of generation.

A successful regulation of transmission access has to deal with the aforementioned difficulties in a satisfactory way. Many of the standing objections to transmission access, and also some of the proposed solutions, fail to integrate the basic economic and power engineering principles that jointly determine the behavior of an electricity market. In the absence of such an integration, these objections and solutions are simply meaningless.

This chapter presents a conceptual (it ignores most implementation issues) regulatory framework of transmission access that satisfactorily handles the seemingly conflicting issues of short term economic efficiency, security of supply, autonomy of individual power systems and financial risk reduction, both in a single or a multiple utility setting. The method combines the use of physically meaningful operation rules and a reduced set of prototype contracts for generation and transmission services.

The emphasis of this chapter is on large power systems consisting of interconnected but independently dispatched entities under competitive regulatory schemes. The structure of ownership and coordination of these entities is diverse. They include single or several vertically integrated utilities; separate distribution, transmission and generation entities (including independent generators); and a power pool with some level of coordination. In the situation considered in this chapter, each one of these entities is assumed to be internally under a single dispatch mechanism of some kind. They will be termed *integrated electric systems* (IESs).

The regulatory framework that is proposed in this chapter interprets and develops the transmission access scheme that is generically known as *third-party access* (TPA). This is the scheme that, in general terms, has recently been proposed to the Council by the Commission of the EU. Under TPA, transmission and distribution entities are required to offer access to their networks to certain eligible entities (major industrial consumers, distribution companies under certain conditions), in exchange for reasonable fees, to the extent that transmission or distribution capacity is available; this provision is supposed to enable the eligible consumers to freely choose their electricity suppliers in the EU. This chapter proposes a specific procedure to accomplish the general objectives defined by TPA.

This chapter takes a neutral stand with regard to the broader issue of how to best promote economic efficiency in interconnected power systems, since this decision is very dependent on the specific context, and other considerations besides economic efficiency are typically involved. In case that a regulation requiring TPA is considered to be the most desirable option, this chapter proposes an approach to handle transmission access that appears to have significant advantages of simplicity, feasibility, and efficiency over alternative schemes.

The regulation scheme to be proposed at the multiple integrated system level in Section 10.3 is, in many aspects, an extension of the simpler regulation that may govern a deregulated single integrated electric system (IES); this is the subject of Section 10.2. In both cases the starting point for the proposed regulation will be the direct application of marginal pricing principles to the different transactions that take place in the corresponding electricity marketplace.

10.2. Prototype Electricity Market for a Single Integrated Electric System

The coordination scheme that will be proposed in this section for a single integrated energy system, and later in Section 10.3 for multiple integrated systems, may in principle be applied to diverse structures of ownership and organization of the entities that comprise each one of the IESs. In this way, most of the autonomy and the characteristics that are particular to each IES may be preserved. However, a strict application of the method would only be adequate when an underlying fully competitive regulatory scheme has been accepted. For the sake of clarity, in this section devoted to the single integrated system level, only a highly deregulated electricity market will be contemplated, since this is the best prototype to show the basic regulation principles that will be used throughout the chapter.

10.2.1. Market Participants

The participants in the deregulated market that is considered in this section are

1. *Generation entities (both utilities and independent generators)* The supply of generation services is deregulated and therefore open to competition. All the generating units are dispatched by a coordination entity that tries to minimize the short-term operation cost of the entire IES.
2. *Distribution utilities* Each utility provides two separate distribution and supply functions. The *distribution* function consists of running the distribution network that carries power from the transmission grid to individual customers, and it is regulated as a natural monopoly, with territorial franchise and regulated tariffs. The *supply* function is in charge of purchasing electricity in bulk to the generators

and selling it to the consumers (it acts as a price insurance entity, basically), and it is deregulated.

3. *Transmission network*   A single entity is in charge of developing, maintaining, and operating the transmission network. Given the natural monopoly characteristics of the transmission function, some kind of rate of return regulation as a public service is adequate, even in a highly deregulated framework (see Bernstein, 1988). However, a carefully designed regulation as a private business is also possible, as in the United Kingdom. The functions of coordination of the dispatching of generation and the provision of secondary network services will be most naturally provided as separate businesses of the transmission entity. Here, for the sake of clarity, they have been listed separately.

4. *Coordination entity*   With this entity, at least the function of coordination of generation dispatch must be provided. Other possible coordination functions that increase the level of integration and economic efficiency within the IES may include from unit commitment to expansion planning (Joskow and Schmalensee, 1983). The coordination of generation dispatch may be organized in different ways, ranging from a pure brokerage system where the coordination entity merely provides the information required to establish the power exchanges, to a completely centralized dispatch based either on costs or bidding prices, including schemes where the coordinating entity may actually buy and sell electricity. Centralized dispatch will be assumed here. It is important to realize (because of the implications for the multiple integrated systems case) that, under conditions of perfect information about the market, the two following situations are equivalent: centralized dispatch of generation and sending spot prices to each generator, which decides how to operate so that it can optimize its profit. For a complete presentation of spot pricing of electricity, see Schweppe, Caramanis, Tabors, and Bohn (1988) and Caramanis, Bohn, and Schweppe (1982). The spot price $\rho_k(t)$ at a given instant of time t and at a given node k is defined as the short-term marginal cost of electricity production with respect to a change in the demand in this node and at this instant of time.

5. *Consumers*   It is assumed that large consumers (in terms of contracted power, energy consumed, or percentage of the IES's total demand) have the right to transmission access and consequently to select their supplier, while small consumers are subject to regulated tariffs and do not have these rights.

10.2.2. Market Transactions

The electricity market is defined by the participants and also by the transactions that are established between them. It is claimed here that, in conceptual terms, the different kinds of market transactions that are commonly established between the participants in an electricity market can be reduced to just three basic prototypic transactions:

1. *Short-term economy transactions* These transactions are directly based on the short-term *nodal* marginal costs (the spot prices) that are "seen" by each participant. Therefore it is not necessary to think in terms of *bilateral* buying and selling transactions in the short-term, since the physical reality of the system operation is totally determined by the centralized generation dispatch and the consumption patterns of the consumers (which may be influenced by the spot prices, if available to them).

The coordination entity makes sure that each participant at the bulk power system level pays or is paid at the corresponding spot price of electricity (see Figure 10.1). A generator that supplies a power g_k at a node k is paid by the coordination entity $\rho_k.g_k$. Any participant (large consumer or distribution utility) that withdraws a power d_k from a node k pays $\rho_k.d_k$. The net revenues of the transmission network are

$$NRT = \sum_k (d_k - g_k) \cdot \rho_k. \tag{10.1}$$

The spot prices ρ_k at bulk system level take into account the marginal costs of generation, transmission losses, and also the out-of-merit and unavailability costs due to the lack of sufficient transmission and/or generation available capacity; all of them associated to the specific node k and instant of time t.

The final consumers connected to the distribution network pay also $\rho_k.d_k$, where ρ_k must now include the extra regulated cost of the distribution network. In order to shield the final consumers (particularly the small consumers) from price fluctuations, regulated stable tariffs may be established by the distribution utilities, that correspond to estimates of the expected values of ρ_k over seasons or entire years. There is no loss of efficiency if the consumers do not have the information or the facilities to respond to price changes in real time.

It is important to realize that, under the above conditions, each generator and consumer is taking the maximum possible advantage of the

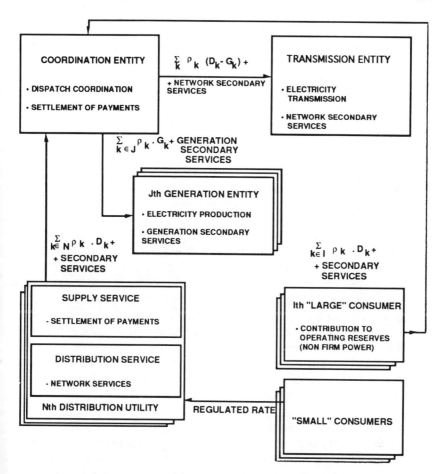

Figure 10.1.   Short-term economy transactions and secondary services

economies of scale of coordination within the IES in the short-term (i.e., the time span that is required to determine the physical operation of the system). Each consumer (including distribution utilities at bulk level) is purchasing electricity at the minimum possible price that is compatible with maximum efficiency of the IES; analogously, each generator is selling its output at the maximum possible prize; therefore there is no justification for generators and consumers to embark in bilateral transactions because there is no additional benefit to be achieved and global efficiency losses will occur, see Schweppe (1988a).

In conclusion, in an IES operating optimally in the short-run according to the proposed regulation, the participants with transmission access rights have implicitly made all possible use of them and the generators must not be allowed to deviate (in their best interest) from the dispatch program that is set by the coordination entity. By definition, the potential security problems that could result from this regulation are nonexistent, since the system is centrally dispatched, with all the security considerations being accounted for in the usual way.

2. *Long-term price insurance contracts* On top of the physically meaningful short-term transactions, purely financial transactions can be established with the objective of reducing the risk associated to the uncertainty of short-term prices (see Figure 10.2). Investors in new generating groups and large consumers of electricity will seek the means of ensuring that prices will not fall or rise beyond certain prescribed limits. These transactions will be purely financial (for a discussion on financial and physical transactions, see Joskow and Schmalensee, 1983), and they will amount to an insurance contract guaranteeing the price of a volume of product for a period of time, and for a fee. Contracts of this kind are presently in place in the U.K., Chilean, and Argentinan systems, for instance. Moreover, it is possible (and it is being done in the United Kingdom) to create a secondary or futures market based on standardized long- and medium-term price insurance contracts.

The proposed format of the price insurance generation contracts is the following (in this example only the customer's purchasing price is insured; the extension to insurance of the generator's selling price or of both simultaneously is trivial): Let p be the upper limit of the spot price ρ_k that a consumer at a node k wants to guarantee for a purchase of a uniform amount of power D^0 during a period T, with f being the per unit fee of the insurance (p and D^0 could be allowed to vary with time of day, season, etc). The insurance firm will typically be a generating entity, but theoretically it could be any other company. Then besides the amount $\rho_k.d_k$ that the consumer must always pay because of the short-term economy transaction, the consumer will pay per unit time

$$f.D^0, \text{ if } \rho_k < p, \text{ and}$$

$$f.D^0 - (\rho_k - p).D_0, \text{ if } \rho_k > p. \tag{10.2}$$

Therefore, this contract can be seen as a *pure bid* on the future evolution of the price ρ_k. It can be signed by a distribution utility with the purpose

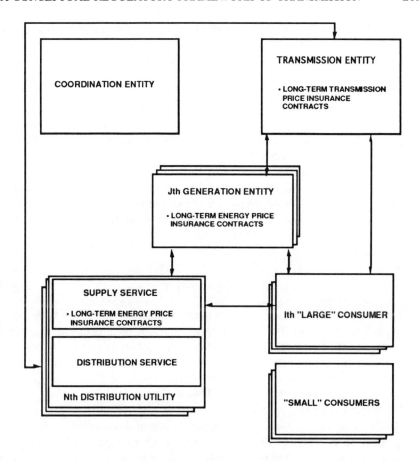

Figure 10.2. Long-term price insurance contracts

of hedging against potential high electricity prices at k, but also by any other firm. It does not interfere with the efficiency of the short-term operation of the system, and it does not have any effect on the security of the power system. It introduces an extra cost for the system as a whole: the insurance fee or risk premium of the contract, that may increase the final cost of electricity.

Do these long-term contracts have any link with the physical power system? They do, but in a form that does not interfere with short-term operation. The payments of price insurance to the customer k happen when ρ_k is high and therefore the prices paid to the generators within the IES (and with more certainty those in the vicinity of k) are also high.

Therefore the entities that are in the best position to offer a price insurance contract to customer k are the generators whose selling prices will approximately follow the evolution of ρ_k during the period of the contract. A generating entity that ensures prices to customers for a volume superior to its expected output is incurring in an additional risk, as well as any other insurance entity whose revenues are not linked to the evolution of r_k. It is therefore clear that physical aspects of the power system, such as unit and fuel availability, maintenance periods, and network transfer capability are relevant to this long-term contract business.

If the payments of the long- and the short-term contracts are jointly accounted for, the total amount that the consumer must pay is (D^0 and d_k are unrelated, in principle, although the consumer will typically contract an insurance for a value D^0 that is an estimate of the future actual demand d_k):

$$f.D^0 + \rho_k.d_k, \text{ if } \rho_k < p, \text{ and}$$

$$f.D^0 + p.D^0 + \rho_k.(d_k - D^0) \text{ if } \rho_k > p. \tag{10.3}$$

Inspection of (10.3) shows that a generator that provides long-term price insurance contracts incurs as a whole in financial risks because of several reasons: unexpected rises in its own variable costs that reduce its margins in the spot price market; underestimation of p; and lower than expected availability that leaves it exposed to the full risk of (10.2) without the coverage of the spot price market. This is the reason of the fee f.

With this prototype contract each node is treated individually, according to its particular ρ_k. Depending on the details of the regulatory system, it may seem more convenient to decompose each spot price ρ_k into a generation component γ that is common to all nodes and a node-dependent network component η_k. Then two parallel markets appear: on one hand the *generation market*, based on the systemwide uniform but temporarily varying γ (a precise definition of γ is not a trivial matter; see Rivier and Pérez-Arriaga, 1993). In a first approximation it can be seen as the marginal cost of generation, resulting in the same short and long term transactions described before, but now replacing ρ_k by γ. On the other hand, the *network market*, is based on the node-dependent per unit prices η_k.

Since in the short-term the system is centrally dispatched, the economic signal η_k to generators and customers without response capability can be averaged over long time intervals without any efficiency loss; however, the remaining customers can make efficient use of it, as with ρ_k before. In the short term, any network user withdrawing a power d_k from the network at node k will pay to the network $\eta_k.d_k$, and a user injecting g_k will pay to the network $-\eta_k.g_k$. Typically η_k will be positive in areas with excess of demand over generation and negative in the opposite case. These are nodal prices and pay no consideration to the hopeless question about the source of the power being received or the destination of the power being injected.

The concept of long-term price insurance contracts for generation services can be directly extended to *transmission network services*, based on the nodal price η_k. Now let q be the maximum price that the network user is willing to pay for each unit of energy to be withdrawn (or injected; the extension is trivial) from the network at node k during a period T, with P^0 being the maximum power that can be removed at any given instant and h the per unit fee of the insurance (q could vary with time of day or the season). The most adequate insurance firm will be the transmission entity, for similar reasons to the ones presented before, but theoretically it could be any other firm. Then besides the amount $\eta_k.d_k$ that the network user (a consumer in this case) must always pay because of the short-term economic transaction, the network user will pay per unit time:

$$h.P^0, \text{ if } \eta_k < q, \text{ and}$$

$$h.P^0 - (\eta_k - q).P^0 \text{ if } \eta_k > q. \tag{10.4}$$

This contract can also be seen as a pure bid on the future evolution of the price η_k. Again it does not interfere with the efficiency of the short-term operation of the system, and it does not have any effect on the security of the power system.

If the payments of the long- and the short-term contracts are jointly accounted for, the total amount that the network user must pay is

$$h.P^0 + \eta_k.d_k, \text{ if } \eta_k < q, \text{ and}$$

$$h.P^0 + q.P^0 + \eta_k.(d_k - P^0) \text{ if } \eta_k > q. \tag{10.5}$$

Similar expressions can be obtained for a generator as a user of the network by simply replacing d_k in (10.5) by $-g_k$.

The contract network method Hogan (1990) and the fixed-price fixed-quantity contracts Schweppe, Caramanis, Tabors, and Bohn (1988) and Schweppe (1988b) have many points in common with the proposed approach. However, here the long-term network contracts are directly established between each participant and the transmission entity, without the need for the participant to find a willing partner to negotiate a bilateral transaction, or for the transmission entity to (arbitrarily) decompose the entire use of the network into node-to-node transactions. These contracts are conceptually simple to implement, and they avoid what in Hogan (1990) seem to be two sources of complexity: the need to identify the "saturation" component of the spot prices and the association with investments in transmission capacity. As in Schweppe, Caramanis, Tabors, and Bohn (1988) and Schweppe (1988b), the proposed transmission price insurance contracts are detached from the actual network use (this is left to the short-term transactions) and from the generation contracts. They are also adequate for a futures market.

3. *Secondary services* These typically include the provision of operating generation reserves, reactive power, stability and frequency control, and black-out start capability. Theoretically, some of these might be accounted for by a comprehensive calculation of the spot prices ρ_k. Here only one type of secondary service will be discussed, since it implicitly appears in most long-term market transactions: this is the firm versus nonfirm attribute of an energy or a network contract. The point to be made is that the concept of *interruptibility* or the lack of *firmness* in a generation or a network contract must be contemplated as the provision of a secondary (or ancillary) service, equivalent to an operating reserve, from the part of the consumer.

By default the two preceding types of contracts must be considered to be firm. Then, any participant may negotiate with the coordination entity the service of interruptibility (reduced reliability or firmness; many variations have been used in real contracts) for a price. A *market of secondary services* must be added to the previous spot market and price insurance market. These services could be negotiated either in the short or in the long term, depending on the situation (if in shortage of operating reserves, extra interruptible load may be contracted in the short-term).

In summary, the firmness of a generation or network transaction is completely unrelated to the prices of the spot market or of the price insurance contracts. Firmness is a separate issue that must be negotiated separately as a secondary service that is *provided by the buyer* of the power when it accepts less than the most reliable supply that the IES can actually provide.

10.2.3. Open Issues

The conceptual regulatory framework that has been presented for the IES leaves a number of open issues, which cannot be discussed at length here. The ones that appear to be most significant follow:

- Total generation costs of some units may not be recovered, after application of the three types of transactions. Others may have revenues in excess of their total costs. A careful definition and computation of the unavailability component of the spot price is needed to make sure that there is enough incentive for new investment in generation. This revenue mismatch is a likely outcome in a competitive deregulated generation market, particularly shortly after the regulatory change that made it possible. Transitory regulation may be advisable during a period (that may be long) of adjustment to the new deregulated environment.
- Total network costs may not be recovered, as well. If transmission is regulated as a public service, the regulation must include a revenue reconciliation mechanism to make sure that the total approved and incurred costs are fully recovered (see Schweppe, Caramanis, Tabors, and Bohn, 1988; Bernstein 1988). If the transmission entity is regulated as a private firm, some revenue and reliability constraints, as well as investment and maintenance incentives, will be needed.
- The maximum power that can be injected or retrieved from a node of the network by a participant that is connected to this node must be subject to some previous negotiation, so that the transmission entity may plan its investments properly. The regulation must specify the time schedules to be observed by both parties.
- The complete procedure of computing spot prices, announcing them, actually dispatching the power system, and charging or paying to the participants must be as consistent as possible, although the nonsimultaneity of all these events will necessarily introduce some inconsistencies.
- The complete remuneration of electric power transactions must include both the active and the reactive components of the electric power. Spot-pricing theory has been developed so that it can be applied to any kind of network representation, therefore resulting in separate spot prices for real and reactive power (see Schweppe, Caramanis, Tabor, and Bohn, 1988). The latter may be useful in providing economic incentives to generators and to distribution companies, regarding their contribution to reactive power and voltage support.

10.3. Prototype Electricity Market for Multiple Integrated Systems

The basic regulation for the multiple integrated systems case results directly as an extension of the rules of the single IES case: When each profit maximizing generator within a single IES receives a price signal that is equal to the corresponding spot price, ideally it operates independently in such a way that the resulting operating point is optimum under the economic efficiency viewpoint of the IES. In the same way, if in a multiple integrated systems setting each integrated system "speaks" with its neighbor in terms of the spot prices at their common borders (they buy and sell energy at the spot price of the specific instant and location), the resulting operation point is the same as the one achieved under a fully centralized dispatch. A simplification of this approach is to consider only one value for the spot price of each integrated system, rather than one per interconnection.

The key piece of regulation that is needed to achieve the above result refers to the coordination entities of all the involved IESs: When dispatching the generation within its own IES, each coordination entity *must not discriminate* between its own generators and the power offered by the neighboring IESs through the interconnections, except for economic reasons. The usual security practices will be used when programming this dispatch. The mechanics of the interaction between the several IESs does not need any particular means beyond what is already available in most IESs energy control centers (see Section 10.3.2 below and Ambrose, 1991; Paula, 1992; and Conejo, Rivier, and Pérez-Arriaga, 1992).

Under the above conditions, each IES as a whole is taking the maximum possible advantage of the existing operational economies of scale of the complete electricity market. This is also the case for each participant within each IES if the IES is regulated as in the prototype electricity market of Section 10.2. If all these prerequisites are met, then the complete regulation system is in full compliance with the fundamental requirements of full access to the transmission network for all participants. The set of open issues in Section 10.2.3 must still be considered for each IES under the prototype regulation. The case of individual IESs with other regulatory schemes will be discussed in Section 10.3.3.

Under the proposed approach, it is not required that the participants engage in bilateral transit (wheeling) transactions. As in the single IES case, these transactions are totally superfluous in a market that is organized and operated according to the principles stated above (see Schweppe, Caramanis, Tabors, and Bohn, 1988; Schweppe, 1988a).

It must be realized that bilateral power transit transactions between participants located in two different IESs (such as a large consumer in the integrated system A that wants to "buy" power from an independent generator in the integrated system B) do not make physical sense until it is precisely defined how they are going to take place. The definition at least requires the establishment of a mechanism that can make compatible the financial contracts of power exchanges with the actual available capacity of the transmission networks, not to mention the operation security constraints and the economic optimality of the dispatch of each IES and also of the entire system. This is frequently ignored in statements about TPA, with the necessary consequence that they are meaningless.

10.3.1. Market Participants

The prototype market considered here consists of any number of IESs that are interconnected but that are also independently dispatched. As pointed out before, a key aspect of the proposed approach is that each IES has its independent coordination entity, so that they can interact on a basis of equality. Therefore, the proposed regulation makes full sense in an environment such as the EU internal electricity market, where the IESs are national power systems or large vertically integrated utilities. This regulatory approach to transmission access will not be directly applicable in cases (such as the U.S. power system) with a large structural diversity and size of the participating entities, a complex pattern of ownership of the transmission network, and lack of a fully encompassing set of IESs (Current Operational Problems Working Group, 1991; Adamson et al., 1991).

Moreover, strict application of the proposed regulation, although beneficial for the economic efficiency of the entire power system, in general will not be in the best interest of each individual participant within each IES. Negotiated consensus, political guidelines, and transitory regulations to deal with particular problems will be needed to achieve the full level of transmission access that is implicit in the proposed regulation.

10.3.2. Market Transactions

The same three types of prototypic transactions can be defined at this higher level of integration:

1. *Short-term economy transactions* These are the power exchanges that take place at the interconnections between neighboring IESs as a result of the nondiscriminatory and independent dispatches that are performed by the coordination entities, based on the respective spot prices. In broad terms, the mechanism that governs these economic

exchanges may be the following: each IES bids prices and quantities for exchanges with the neighboring IESs (only physical transactions are of interest here); the coordination entity of each IES prepares the exchange program as part of its economy/security dispatch function; or the standard load-frequency control procedure can now be followed. There is no reason why this procedure may deteriorate the level of security of the system. As it was mentioned before, there is no need (as it would be harmful for the economic efficiency of the entire system and no participant could benefit from it) for any additional bilateral short-term power transits between noncontiguous IESs or between non-IES entities that belong to either the same or different IESs.

2. *Long-term price insurance contracts*   The situation for these contracts is very much the same as in the IES case. Now the entities from an IES can make price insurance contracts of generation or network services with entities located in a different IES. The only additional point to be made is that, for the same reasons presented in Section 10.2.2, a generator that is located close to a consumer is in a better position (at smaller risk) to provide a price insurance than a generator in another IES, where the pattern of spot prices may be quite different, particularly if there are active transmission constraints in between. The coordination entities of the IES may sign insurance contracts between themselves or with other firms, concerning the prices of the power being exchanged at the interconnections.

3. *Secondary services*   The IESs between themselves can agree on the provision of services such as operating reserves, black-out start support, or stability. The same considerations that were made in Section 10.2.2 concerning firm versus nonfirm power are applicable here.

10.3.3. Open Issues

The following issues deserve a detailed discussion that cannot be provided within the scope of this chapter:

- Nondiscriminatory dispatch may result in severe cases of stranded generation investment in those IESs that must increase significantly their energy imports because of the proposed regulation. Transitory regulation may be needed to mitigate these situations.
- There is no reason to believe that there is a special opportunity for the IESs to exert market power in a multiple integrated systems

environment. The potential of an IES for wielding market power in a neighboring integrated system will depend on the value of the transfer capability of the interconnection between both, relative to the demand of the neighboring system.

- The implementation of the proposed transmission access regulation scheme at the multiple IES level, does not necessarily require the existence of a spot-price-based energy marketplace in the participating integrated systems, as in Section 10.2. However, it does require that the participants with transmission access rights (either generators or consumers) have to be under spot prices. This may result in disadvantage for the remaining participants within the considered IES, if the participants with transmission rights pay less (or are paid more) than their share under the current IES regulation. Therefore, although the autonomy of ownership, organization and operation (subject to the non-discriminatory dispatch rule) of each IES can be preserved, the IESs may have to implement transitory (or even permanent) regulations to keep the participants without transmission rights from being disadvantaged. One possible approach may consist of demanding economic compensations from the participants that choose to switch from the present regulatory scheme to the one that is presented here.

- Under strict conditions of nondiscriminatory dispatch of every IES, the spot prices at any border node between neighboring IESs will be equal when seen from both sides, even when tie lines are saturated, therefore discouraging further trade. Using distinct spot prices at each one of the frontier nodes, rather than a single value per IES, will increase the precision but also the complexity of the procedure that is needed to reach the equilibrium at a given time (Conejo, Rivier, and Pérez-Arriaga, 1992). Methods to compute these spot prices are readily available (see Rivier, Pérez-Arriaga, and Luengo, 1990, for instance).

10.4. Conclusions

It is possible to reconcile the requirements of coordination and competition that underlie the proposals of third party access with a satisfactory solution to the issues of short-term economic efficiency, security of supply, autonomy of individual power systems, and financial risk reduction. This chapter has presented a conceptual regulatory framework that achieves these objectives via the combined use of physically meaningful operation rules and a limited set of prototype contracts for generation and transmission services. The proposed regulation is applicable at both single and multiple integrated system levels. A number of relevant open issues have been identified at both levels; they need to be solved for specific power systems at the next and more detailed step in the development of the regulation.

Acknowledgments

Most of the ideas presented in the paper have been developed during the realization of studies supported by Red Eléctrica de España, ENDESA e Iberdrola. The author is also thankful to Carmen Illán, Michel Rivier and Guadalupe de Cuadra for helpful discussions.

References

Adamson, A.M., L.L. Garver, J.N. Maughn, P.J. Palermo, and W.L. Stillinger. 1991. "Summary of Panel: Long-Term Impact of Third-Party Transmission Use." Paper presenteda at the IEEE 1991 PES Winter Meeting, paper 91 WM 007–5 PWRS.

Ambrose, D.R. 1991. "Inter-Utility Communications Within WSCC." *IEEE Transactions on Power Systems*, 6(4) (November).

Bernstein, S. 1988. "Competition, Marginal Cost Tariffs and Spot Pricing in the Chilean Electric Power Sector." *Energy Policy* (August).

Caramanis, M. R.E. Bohn, and F.C. Schweppe. 1982. "Optimal Spot Pricing: Practice and Theory." *IEEE Transactions on Power Apparatus and Systems*, vol. PAS-101, no. 9 (September).

Commission des Communautés Européennes. 1991. "Rapports des Comités Consultatifs sur l'Accès de Tiers aux Réseaux Electriques." *Direction General de l'Energie* (May).

Conejo, A., M. Rivier, and I.J. Pérez-Arriaga. 1992. "Application of the Dantzig-Wolfe Decomposition to Decentralized Optimal Dispatch of Interconnected Electric Power Systems." Paper presented at the 20th Reunión Nacional de Estadística e Investigación Operativa, September (in Spanish).

Current Operational Problems Working Group. 1991. "Operating Problems with Parallel Flows." *IEEE Committee Report, IEEE Transactions on Power Systems*, 6(3): 1024–1034.

Edison Electric Institute. 1986. "The Vital Link: Electric Transmission and the Public Interest."

Hogan, W.W. 1990. "Contract Networks for Electric Power Transmission." 1990. Kennedy School of Government, Harvard University, September.

Joskow, P.L., and R. Schmalensee. 1983. *Markets for Power*. Cambridge, Mass.: MIT Press.

Paul, G. 1992. "Network Security: System Exchanges EMS Information." *Electrical World* (February).

Rivier, M., and I.J. Pérez-Arriaga. 1993. "Computation and Decomposition of Spot Prices for Transmission Pricing." Paper presented at the 11th Power Systems Computation Conference, Avignon, France, September.

Rivier, M., I.J. Pérez-Arriaga, and J. Luengo. 1990. "JUANAC: A Model for Computation of Spot Prices in Interconnected Power Systems." Paper presented at the 10th PSCC Conference, Graz, Austria, August.

Schweppe, F.C. 1988a. "Mandatory Wheeling: A Framework for Discussion." Paper presented at the IEEE 1988 PES Summer Power Meeting, paper 88 SM 690–0, July.

———. 1988b. "A Spot Price Based Transmission Marketplace." Unpublished paper.

Schweppe, F.C., M. Caramanis, R.D. Tabors, and R.E. Bohn. 1988. *Spot Pricing of Electricity*. Norwell, Mass.: Kluwer.

11 COMPETITIVE JOINT VENTURES AND ELECTRICITY TRANSMISSION

Susan P. Braman[1]

Given its natural monopoly status, electricity transmission is achieved at lowest cost with only one set of facilities. Nonetheless, joint ownership of these facilities has the potential to create a competitive environment and thereby contribute to efficiency in the supply of transmission services. Under a joint ownership structure, competition comes from competing marketers of output from a single set of facilities in which those marketers own shares of capacity. Furthermore, to prevent the owner-marketers from restricting capacity, a competitive joint venture (CJV) requires open entry into ownership—that is, anyone is allowed to become an owner by paying for expansion of capacity. Put simply, a CJV establishes competition in both the short-run and long-run supply of output.[2]

Competitive joint ventures provide a market solution to the problem of natural monopoly: minimum-cost production is achieved by having a single plant, but competitive outcomes are implied by open entry and independent marketing. Price and output need not be regulated, and pricing flexibility is afforded. Price flexibility is particularly valuable for services like electricity transmission, where demand often fluctuates randomly. In addition to providing a tool for regulatory and antitrust policy, competitive joint ventures have potential use in the context of privatization.

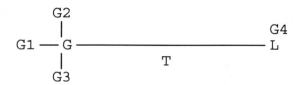

Figure 11.1. A Simple Electric Power Delivery System

The next section lays out a two-stage model of electricity transmission structured as a competitive joint venture. The model incorporates decreasing long-run average cost, increasing short-run marginal cost and random demand. However, it abstracts from the network aspects of electricity transmission by considering a situation in which transportation of electricity from point A to point B is desired and there is only one path between the points.[3]

In Section 11.2, I describe the short-run pricing of transmission services under CJV ownership. I assume that there is central dispatch of generated electric power and that short-run prices for transmission are determined as a byproduct of this dispatch. This is simply an application of Hogan's (1992) contract network scheme, except that in a CJV, capacity rights are obtained through ownership rather than through contract. There is short-run open access to the transmission grid, efficient usage prices, and compensation to owners of capacity. In the long run, then, the stream of short-run costs and revenues can be anticipated and used as a basis for individual capacity choices. Section 11.3 describes the long-run equilibrium in a CJV with open entry. Welfare implications are discussed in Section 11.4 and Section 11.5 concludes.

11.1. Market Conditions

Electric power delivery systems consist of three distinct vertical stages: generation, long-distance transmission, and local distribution. I assume that generation (the wholesale power market) is potentially competitive and that the market for transmission services is a natural monopoly. Issues in distribution services are not considered. Consumption of power is assumed to take place at the off-take node; one can think of the buyer market as consisting of a single price-taking industrial customer.

As an example, Figure 11.1 illustrates the type of situation that I am considering. Generators G1, G2, and G3 compete to supply power at node

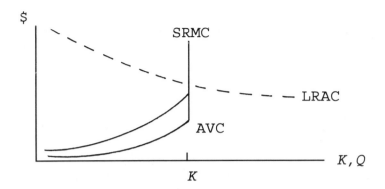

Figure 11.2.   Two-Stage Model

G. Generator G4 is adjacent to the load (local distributor or customer) at node L but is a higher-cost source of power than are G1–3. The transmission line T connects node G with node L and is the focus of our interest. The "transmission problem" can be illustrated here by supposing that generator G1 owns T and is reluctant to provide transmission service to G2 or G3 because this would allow them to compete with G1 to supply L.[4] The solution that I am proposing to this problem is that T would be structured as a CJV.

Consider the following two-stage model of electricity transmission over a single line. Capacity level K is chosen in stage 1 and transportation of $Q \leq K$ units then occurs in stage 2. The production technology exhibits economies of scale in the capacity to transport but not in actual transportation from existing capacity. In the (long-run) first stage, capacity is acquired at decreasing average cost. Once a capacity level is chosen, its cost is fixed and the original long-run average cost (LRAC) curve is no longer relevant; a different cost function will govern subsequent expansion. In the (short-run) second stage, the capacity level is fixed and the marginal cost of transportation increases gradually up to the capacity constraint, at which point it becomes infinite. Such a cost structure results in a natural monopoly since the service is produced at lowest average cost by a single plant. In Figure 11.2, SRMC is the short-run marginal cost, and AVC is the average variable cost, of transportation, given capacity level K.

Stage 2 is a series of short-run periods in which ownership shares and total capacity are fixed. Since demand for transmission services is derived from the demand for delivered power, it is assumed to fluctuate randomly in the short run. An inverse short-run transmission demand function is denoted by $p(Q;\theta)$ where Q is the total quantity of power transported and θ is a realization of a random variable θ with some stable distribution. In each stage 2 period, such a demand function is realized and becomes known. In stage 1, potential owners of capacity are assumed to know the distribution of demand.

A CJV consists of an operator who produces output from a single plant or set of facilities and n co-owners who have property rights to their shares of capacity. At least two rules govern the joint venture in order to increase efficiency in both stages of decision-making. In stage 1, anyone is allowed to become an owner by claiming units of capacity and paying for them at the average cost of total capacity. In stage 2, competitive outcomes are facilitated by either a limit on the concentration of ownership or by removing the ability of individual owners to withhold capacity from the market.[5]

11.2. Electricity Transmission CJV: Short-Run Pricing

For an unconstrained transmission line, the major determinant of short-run cost is the energy lost in transportation. The following notation is used:

$$l = l(Q;K) \text{ is total short-run cost (losses)},$$

$$l_1(Q;K) = \frac{\partial l}{\partial Q} = SRMC, \; l_1(Q;K) > 0, \; l_{11}(Q;K) \geq 0,$$

$$\frac{l}{Q} = AVC, \text{ and}$$

$$l_2(Q;K) = \frac{\partial l}{\partial K} \leq 0.$$

When the line becomes constrained, the cost of transmission also includes the opportunity cost of not being able to use the line. If extra capacity is maintained as an emergency buffer, then K is the *effective* capacity to be used generally and the opportunity cost rises when K is reached.

We may now consider various models of short-run competition. If short-run marginal cost were constant, we could suppose that each marketer chooses a price at which to offer its output. Or, in the case of increasing marginal cost without joint use, we could suppose that each marketer of transmission services chooses an offer curve to submit to a central market "auctioneer."

But the sharing of operating cost when marginal cost is increasing creates externalities that yield suboptimal outcomes under these models. First, the marginal cost faced by an individual owner-marketer is below the true marginal cost.[6] Second, the marginal cost faced by an individual owner depends on the use of the line by other owners.[7]

Furthermore, the nonstorability of electric power together with the randomness of demand create a need to balance electricity generation and load in real time. The value of a transmission system derives not only from reducing the cost of delivered power but also from the maintenance of reliability. Central dispatch of generators (described below) is essential to meeting the reliability and efficiency standards of modern electricity delivery systems (Hogan, 1992). Even in our simple example from Figure 11.1, central dispatch is necessary if the load at L is to be served by the least-cost generation plant at each point in time. Thus, the "systems" nature of electricity generation, transmission, and consumption has implications for the extent of decentralization that can be accommodated in the short-run market for transmission services.

Fortunately, however, there is a nonmarket mechanism for achieving efficient pricing of transmission services. It is based on central dispatch of generating plant, a mechanism by which the system operator uses information about generation costs, transmission costs, and demand to serve the system load at minimum possible cost.[8] Each generation unit on the system submits an offer of the prices at which it is willing to supply various amounts of power to the grid at its node.[9] The grid operator then uses this information, along with its knowledge of power flows, grid losses, and grid constraints to find the sequence of dispatching generating plant that will minimize the cost of serving loads. The spot price of energy at each node in the system is then the minimum cost at which another unit of power can be delivered to that node. This price will include the price of generated power, the marginal cost of transmission line losses, and the marginal opportunity cost of transmission constraints (Schweppe, Caramanis, Tabors, and Bohn, 1988).

In our example from Figure 11.1, the operator would meet the increased demand at L by calling for more output by the generator with the lowest marginal cost at that time. Marginal cost information would be reported in advance through price-quantity offer schedules by G1, G2, and G3 to supply power to node G. Generator G4 would offer to supply power to node L. The operator would determine the (minimum) marginal cost of energy at node G and at node L. As long as there is excess capacity in T, the delivered energy price at node L may be the price of power from G1, G2, or G3 plus the cost of transmission losses or it may be the price of power from G4, whichever is less. If T is capacity-constrained, then the minimum marginal cost of energy at L is that of G4.

Conceptually, transmission owners could submit offers, as generators do, of the prices at which they are willing to supply services from their shares of transmission capacity. These would then be used, along with the generator offers and in place of the true grid costs, to determine the optimal sequence of dispatching. The transmission market would then be decentralized. However, the joint use together with increasing marginal cost yields the externality problems noted above. Furthermore, unlike the case of generation, the grid operator actually does know the costs of transmission (indeed, its knowledge is better than that of the owners, moment to moment). Therefore, rather than having the central dispatch be guided by the offers of transmission owners, several authors have suggested that the optimal dispatch (using actual transmission costs) should be taken as given and transmission spot prices calculated ex post of dispatch (see Kelly et al., 1987; Read, 1988; Read and Sell, 1988; Hogan, 1990, 1992; and McCabe et al., 1991). This section only briefly lays out the way in which such a mechanism might work. Greater detail is provided by Hogan (1990, 1992).

Once least-cost dispatch of power has taken place, the value of transmission between two nodes in the network is then the difference between the spot prices of energy at the two nodes. In effect, least-cost dispatch simulates the outcome of a competitive market and efficient prices for transmission between any two nodes can then be determined ex post by differencing the marginal costs of delivered energy at the nodes (Hogan, 1992). In our two-node example, the price for use of T is the difference between the marginal costs of delivered energy at L and at G.

When the line capacity is not constrained (demand is realized as p_1 in Figure 11.3), the efficient price of transmission is simply $l_t(Q^*;K)$, the marginal cost of energy lost in transmission of the quantity Q^*, which equates marginal cost and inverse demand.[10] When the capacity constraint becomes binding (p_2 is realized in Figure 11.3), the shadow price of the constraint is measured by the increase in the marginal cost of energy caused by the inability to make further use of the least-cost generating unit. The cost of this "out-of-merit" dispatch is rightly added to transmission losses in determining transmission prices (Hogan, 1992). This is analogous to any peak demand situation, when price must rise to $p^* = p(K;\theta)$ in order to clear the market.

In our example, suppose that demand increases at L and that the least-cost way to meet the increase is to transport more power from G2. If there is no excess capacity in T, then the increased demand must be met instead by G4. This raises the marginal cost of energy at node L above what it would otherwise be and therefore, the price of transmission also rises. This

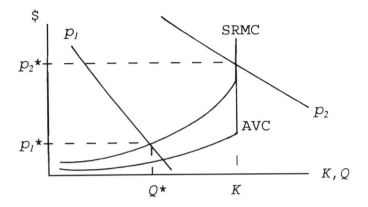

Figure 11.3.   Peak and Off-Peak Pricing

congestion charge for transmission reflects its opportunity cost and acts as a signal for investment decisions.

Whether or not capacity is constrained, the efficient price always yields short-run revenues over and above operating costs since $l_i(Q;K) > l/Q$. These short-run profits offset the cost of capacity and signal the benefits from investing in capacity.

Property rights and trade in a CJV transmission system may operate in the short run as follows. Anyone is allowed unlimited short-run access to the grid subject to the condition that they agree to pay the prices determined by the operator. Suppose that, in our example, G3 is not an owner in the CJV, but that G1 and G2 (and perhaps other entities) are. Any-one—G1, G2, G3, G4, L, or anyone who wishes to use transmission to facilitate a trade—may use the line. The CJV participants (owners of property or "capacity" rights in the line) then receive back the difference between total revenue and actual transmission operating cost in proportion to their ownership shares (Read and Sell, 1988; Hogan, 1992). Holders of capacity rights are never subject to the higher costs, when using their own capacity, caused by a transmission constraint and are therefore "hedged" against the higher costs arising from system constraints.

Reliance on the operator to determine efficient short-run prices has two notable implications. First, concentration of ownership is not a concern because no owner is able to exercise market power in transmission services.[11] Owners of capacity cannot choose to hold idle capacity and the "market" price is the efficient one, determined centrally by the operator.[12]

Second, the incentives for and ability of the operator to become a cartel-enforcement agent must be reckoned with. The owners may be able to maintain the monopoly price and output by having the operator use an inflated variable cost function to determine prices.[13] Some type of government monitoring is required to ensure that the mechanism that determines price is operating as it should.

11.3. Electricity Transmission CJV: Long-Run Equilibrium

Thus far in the discussion, the amount of transmission capacity owned by each participant has been fixed and only the output from that capacity has been flexible. Now we ask about the level of capacity that will be chosen in a CJV with open entry. In our electricity transmission context, the maximum amount of power that can be transported is determined by the capacity of the transmission line.[14] Once this is chosen, it cannot be varied instantaneously and is thus fixed over some stage 2 period.

The open-entry rule is imposed to ensure that if long-run profits are expected to accrue from a given level of capacity, then capacity will be expanded. Whether firms expect long-run profits at a particular anticipated level of total capacity depends on the way that the capacity level affects prices in the short run. Since the short-run market operates efficiently, expected average revenue can be derived from the distribution of inverse demand and the knowledge that the price in each short-run period will be $l_1(Q^*;K)$ or $p(K;Q)$, whichever is greater.

Long-run expected revenue less operating cost, per unit of capacity, is then

$$P(K) = \int_{\underline{\theta}}^{\tau} \frac{Q^* p(Q^*;\theta) - l(Q^*;K)}{K} f(\theta)\, d\theta + \int_{\tau}^{\bar{\theta}} [p(K;\theta) - \frac{l(K;K)}{K}] f(\theta)\, d\theta, \tag{11.1}$$

where $f(\theta)$ is the probability density function for the random component of market demand, $\underline{\theta}$ and $\bar{\theta}$ are the lower and upper bounds on the support of $f(\theta)$, Q^* is the quantity that equates marginal cost and inverse demand (i.e., $l_1(Q^*;K) = p(Q^*;\theta)$) for a given θ, and τ is the value of θ that yields $Q^* = K$ (i.e., $\theta = \tau$ yields an inverse demand function such that capacity is just fully used at a price equal to marginal cost).[15]

This expected value of average net short-run revenue at capacity level K is derived from the fact that, for all realizations of demand such that the line is not constrained ($p(K;\theta) \leq l_1(K;K)$), price is equal to $l_1(Q^*;K)$ and quantity is Q^*; and for all realizations of demand such that a capacity constraint is binding ($p(K;\theta) > l_1(K;K)$), price is equal to $p(K;\theta)$ and quantity is K (refer to Figure 11.3). It is assumed that the integral (11.1) exists and that $P(K)$ is twice-differentiable and strictly decreasing in K.

Denote the per-period cost of capacity to transport Q units of electricity as $C(Q)$ and assume that $C(0) = 0$ and that $C(\cdot)$ is concave, twice-differentiable, and strictly increasing in Q. Then long-run average cost (LRAC) is continuous and strictly decreasing over all values of Q. The measurement of capacity is normalized so that one unit of capacity is capable of transporting one unit of electricity per period, so $C(Q) = C(K)$.[16]

Assume that there is an infinite number of risk-neutral potential owners whose benefit from ownership derives entirely from the expected revenue stream and not from consumer surplus; that is, no potential owner is an end-user of delivered power.[17] Each potential owner i knows $P(K)$ and $C(K)$, chooses k_i units of capacity, and expects long-run profit per period of

$$\pi_i^e(k_i|k_{-i}) = k_i[P(K) - \frac{C(K)}{K}],$$

where k_{-i} is the capacity held by all potential owners other than i, and $K \equiv k_i + k_{-i}$. The expected profit function is assumed to be concave in k_i for fixed k_{-i}.

Suppose that total capacity is determined by an iterative process in which potential entrants first simultaneously choose individual capacity levels. The total capacity level (and thus, average cost) is determined and reported to all players. Players are then allowed to adjust their individual capacity levels by simultaneously choosing new levels. A stage 1 equilibrium is reached when no further adjustments are desired. The total level of capacity is then put into place. Some amount of extra capacity, above the amount that will be normally used, may be required for safety reasons. The amount of this buffer may be chosen jointly by the owners and should be thought of as a technological necessity and not as affecting price.

As shown below, the equilibrium total capacity level in a CJV must be one that makes potential owners indifferent between entering and not entering. Proposition 1 characterizes the stage 1 equilibrium as the zero-profit point, as the number of owners grows large.[18]

Proposition 1: As the number of nonuser owners n approaches infinity, K_n approaches K^*, where K_n is the total capacity chosen by n owners and $P(K^*) = C(K^*)/K^*$.

Proof: We first derive K_n as a function of n. Each individual owner chooses k_i to maximize $\pi_i^e = k_i[P(K) - \frac{C(K)}{K}]$. The first-order condition is

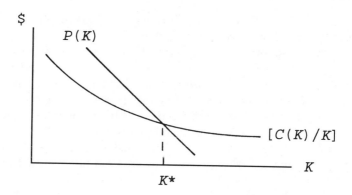

Figure 11.4. Long-Run Equilibrium with Nonuser Owners

$$P(K) + k_i P'(K) = \frac{C'(K)k_i}{K} + \frac{(K - k_i)C(K)}{K^2},$$

so that

$$k_i = \frac{K[C(K) - KP(K)]}{K^2 P'(K) - KC'(K) + C(K)}.$$

The second-order condition is satisfied by concavity of π_i^e. By symmetry of the profit functions, $k_i = k$ for all i, and total capacity is

$$K_n = nk = \frac{nK_n[C(K_n) - K_n P(K_n)]}{K_n^2 P'(K_n) - K_n C'(K_n) + C(K_n)}$$

After algebraic manipulation, this can be written as

$$P(K_n) + \frac{K_n P'(K_n)}{n} = \frac{(n-1)C(K_n)}{n} \frac{C(K_n)}{K_n} + \frac{C'(K_n)}{n}.$$

The limit of this expression, as n goes to infinity, is

$$P(K_n) = \frac{C(K_n)}{K_n}. \qquad\qquad\blacksquare$$

The equilibrium capacity level of Proposition 1 is K^* in Figure 11.4. It is the level of capacity that generates average revenue equal to average cost.

11.4. Welfare Analysis

As long as the dispatch process results in transmission prices equal to short-run marginal cost (including opportunity costs), we know that welfare in stage 2 is maximized. We also know that, with open entry in stage 1, the CJV structure will result in a capacity level that equates average capacity cost and average net revenue from short-run trade at efficient prices. The remaining question is, "What are the welfare implications of this long-run outcome?"

Williamson (1966) shows that the problem of choosing capacity with periodic loads can be reduced to an analysis of "effective demand-for-capacity" or its inverse, the marginal value of capacity.[19] The marginal value of capacity can be derived from the total value of capacity as follows. Let $V(K)$ denote the total-value-of-capacity function (gross of capacity cost). For $\theta \leq \tau, V(K)$ is the surplus generated by transportation at level Q^*, where Q^* equates $p(Q;\theta)$ and $l_1(Q;K)$. For $\theta > \tau$, $V(K)$ is the surplus generated by transporting $Q = K$ units. Therefore,

$$V(K) = \int_{\underline{\theta}}^{\tau}\int_0^{Q^*} [p(Q;\theta) - l_1(Q;K)] \, f(\theta) \, dQ \, d(\theta)$$
$$+ \int_{\tau}^{\bar{\theta}}\int_K^0 [p(Q;\theta) - l_1(Q;K)] \, f(\theta) \, dQ \, d\theta. \qquad (11.2)$$

The marginal value-of-capacity is approximated as follows, where I ignore second-order effects (that are due to the fact that both Q^* and τ are functions of K because they depend on $l_1(Q;K)$).

$$V'(K) \approx M(K) = -\int_{\underline{\theta}}^{\tau} l_2(Q^*;K) \, f(\theta) \, d\theta +$$
$$+ \int_{\tau}^{\bar{\theta}} [p(K;\theta) - l_1(K;K) - l_2(K;K)] \, f(\theta) \, d\theta \qquad (11.3)$$

Intuitively, for a given realization of demand such that $\theta > \tau$ (the second term in (11.3)), the marginal willingness to pay for capacity is the difference between the market-clearing price and marginal cost at K *plus* the change in total variable cost due to the cost curve shifting down ($l_2(Q;K) \leq 0$). For $\theta \leq \tau$, the marginal willingness to pay for capacity is only this change in total variable cost (evaluated at Q^*) since in this region, increases in K do not result in discrete increases in surplus from increased output (see Figure 11.5).

Optimality requires that K be chosen to equate $V'(K)$ and $C'(K)$. The welfare implications of the CJV capacity choice therefore depend (approximately) on the relationship between $M(K)$ and $P(K)$. To analyze

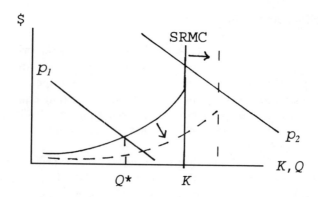

Figure 11.5. The Marginal Value of Capacity

this relationship, consider the following special cases. Note first that if $l(Q;K) = l(Q) = cQ$, where c is a constant, then variable cost does not depend on K $(l_2(Q;K) = 0)$ and short-run marginal cost is constant $(l_1(Q;K) = l/Q = c)$. In this case, $M(K) = P(K)$ and open entry into the CJV results in a second-best level of capacity where the marginal value of service is equal to long-run average cost. This characterization may be used as an approximation of the outcome, since the cost of energy losses in electricity transmission is fairly flat until the quantity transported approaches closely the capacity of the line. The upward slope of SRMC as Q increases, and the downward shift of SRMC as K increases, probably occur only in the neighborhood around the existing K.

When short-run costs do depend on the level of capacity chosen $(l_2(Q;K) < 0)$, but these costs are constant per unit $(l_1(Q;K) = l/Q)$, then $P(K)$ is less than $M(K)$ for all K. In this case, the level of capacity chosen will be less than that which equates marginal value and average cost.

When short-run marginal cost is increasing $(l_1(Q;K) > l/Q)$, but does not depend on the level of capacity $(l_2(Q;K) = 0)$, then $P(K)$ is greater than $M(K)$ for all K. Here, the level of capacity chosen under a CJV is larger than that which equates marginal value and average cost.

In the model of electricity transmission, both complications appear (short-run marginal cost is not constant and it decreases with capacity), and the relationship between average revenue $(P(K))$ and marginal value $(M(K))$ will depend on K. In reality, however, short-run marginal transmission costs may be approximately constant until Q is very close to K, so that a good approximation is achieved with the case in which neither complication is present.

11.5. Conclusions

For policy makers, a competitive joint venture offers an approach to natural monopolies that does not require that a regulator have a lot of information about costs or demand. In the case of electricity transmission, the regulator may need to verify the short-run operating costs (electricity losses) that are used in the dispatch process, but otherwise, no knowledge is required.

Central economic dispatch, by a grid operator who knows transmission costs and collects offers to supply generated electricity, results in a first-best short-run allocation of capacity. Use of transmission can then be valued ex post and priced at its marginal cost. If the resulting short-run profits are returned to the owners of transmission capacity, this revenue stream yields an incentive to invest in capacity. Requiring open entry into a joint venture results in a long-run level of capacity that approximately equates marginal value and average cost. This outcome enables cost recovery with no regulation of rates of return and no price discrimination in a given demand state.

An unresolved problem is that of the potential moral hazard created by the principle-agent relationship between the CJV owners and the grid operator (supplier of production services) or construction contractor (supplier of capacity services). These agents may be able to profit by failing to minimize costs, and individual owners have little incentive to monitor the agents' behavior. One source of inefficiency in monitoring is that any cost reduction will be available to all owners: this problem is the same as that in a corporation where owners are dispersed and have little individual incentive to monitor management. Another source of inefficiency is specific to CJVs: even as a group, the owners cannot gain by reducing long-run costs since the competitive environment in a CJV results in the long-run average price being reduced as well. The CJV owners do have incentive to minimize costs in the long run if they face potential competition from new generators, however. Furthermore, in contrast to behavior under rate-of-return regulation, CJV owners themselves have no incentive to raise investment costs ("gold plate"), since their profit is not tied administratively to the amount of investment.

Notes

1. The views expressed are those of the author and do not necessarily reflect those of the Federal Trade Commission or of any individual commissioner.
2. The rules governing a CJV are based on those suggested by Vernon Smith (1988).

3. I believe the model to be generalizable to more complicated networks, but here I wish to keep things as simple as possible in order to illustrate key points. Electric transmission networks generally involve externalities in both use and investment, and models are necessarily complex. Fundamentally, however, I believe the joint-venture structure holds more promise for dealing with externalities than do other ownership or institutional arrangements.

4. Theory predicts that, if transmission rates are not held down by regulation, service would never be withheld but would be priced so as to remove the gain to competitors from encroaching on the market. This does not, however, solve the problem.

5. The second option is known as a "use-or-lose" rule and fundamentally alters an owner's property right so that it may sell output only at the efficient price. As will be discussed, in electricity transmission, a use-or-lose rule is feasible because the efficient price is easily derived from the central dispatch mechanism. In other applications, however, it is expected that the rule would be a limit on the concentration of ownership in the joint venture. The problem with the concentration limit from a theoretical standpoint is that the individual capacity constraints may prevent achievement of competitive outcomes when the number of owners is finite. For basically the reasons described by Edgeworth (1925) in his critique of Bertrand price competition, equilibria will be in mixed strategies and characterized by withholding of output from the market. The use-or-lose rule removes individual capacity constraints—the source of market power in Edgeworth's story—and thus, has the same effect as an infinite number of owners. In practice, the concentration limit can be used to arbitrarily approach short-run efficiency. See Braman (1992) for analysis of short-run equilibria in a CJV.

6. True total short-run cost can be written as $l = Q\dfrac{l}{Q}$, so true marginal cost is

$l_1 = \dfrac{l}{Q} + Q\,[\dfrac{Ql_1 - l}{Q^2}]$. Compare this to the situation faced by an individual i, Whose total short-

run cost is $l_i = q_i\dfrac{l}{Q}$, where q_i is the quantity produced by that individual $(\sum_j q_j \equiv Q)$, and

marginal cost is $\dfrac{\partial l_i}{\partial q} = \dfrac{l}{Q} + q_i[\dfrac{Ql_1 - l}{Q^2}]$.

7. In more complex networks, another cost externality is that the use of one transmission line cannot be separated from the use of other lines with which it is connected. Often referred to as "loop flow," the fact that power flows along the path of least resistance means that the cost of a particular efficient transaction cannot always be isolated; an action taken in one part of the grid may affect the costs incurred at remote locations by those not party to the transaction (Federal Energy Regulatory Commission, 1989).

8. Central dispatch is also referred to as "least-cost" or "economic" dispatch.

9. Ideally, these are marginal cost prices; however, the level of competition in the generation market will affect prices.

10. That is, $l_t(Q^*;K) = p(Q^*;\theta))$ for a given θ.

11. Read and Sell (1988) note that market power *is* a concern when some player is able to affect the spot price of *power* at a node and thereby affect the price of transmission.

12. There is in effect a "use-or-lose" condition.

13. For example, suppose that true total operating cost is $l = .2Q^2$, so that short-run marginal cost is $.4Q$. Suppose further that realized inverse demand is $p(Q) = 100 - .4Q$. Then the efficient price and quantity is determined by $p(Q) = .4Q$, or $Q^* = 125$ and $p^* = 50$. In contrast, the joint profit-maximizing price and quantity are determined by setting marginal revenue equal to marginal cost, or $100 - .8Q = .4Q$, which yields $Q^m = 83$ and $p^m = 67$. Note

that if the operator uses the function $MC^m = .8Q$ (rather than $.4Q$) to represent short-run marginal cost, then the price determined by the dispatch process will be p^m.

14. In a complex network, the capacity of any one line is a function of the grid configuration, as well as the capacity rating of the line itself.

15. τ is the value of θ that is the cut-off between peak and off-peak demand, or constrained and unconstrained line use.

16. Note that I have assumed that cost depends only on the capacity, and not the length, of the line. In our example, points G and L are fixed, so $C(Q)$ is the cost of transporting Q units from G to L. In reality, investment choices may be multi-dimensional.

17. In the example, the consumer at L is not a potential owner of transmission. See Braman (1992) for an analysis of the issues involved in ownership by end-users.

18. Gale (1992) derives this result for the case in which demand is fixed over short-run periods and short-run marginal cost is constant.

19. Williamson (1966) analyzes the problem of choosing capacity when demand is periodic but deterministic. The analysis is equally applicable to random demand, since the same capacity will be used whatever the realization of demand. The crucial point is that the demand curves are summed vertically, not horizontally, because the use of capacity is nonrival over short-run periods.

References

Braman, S. 1992. "Theory and Application of Competitive Joint Ventures." Ph.D. dissertation, Georgetown University.

Edgeworth, F. 1925. "The Pure Theory of Monopoly." In *Papers Relating to Political Economy*, (vol. 1). New York: Macmillan.

Federal Energy Regulatory Commission. 1989. *The Transmission Task Force's Report to the Commission. Electricity Transmission: Realities, Theory, and Policy Alternatives*. Washington, DC: FERC.

Gale, I. 1992. "Price Competition in Noncooperative Joint Ventures." Mimeo, University of Wisconsin.

Hogan, W. 1990. "Transmission Pricing in New Zealand" and "Transmission Capacity Allocation and Charges." Papers from Putnam, Hayes and Bartlett prepared for the Electricity Corporation of New Zealand.

———. 1992. "Contract Networks for Electric Power Transmission." *Journal of Regulatory Economics*, 4: 211–242.

Kelly, K., J.S. Henderson, and P. Nagler. 1987. "Some Economic Principles for Pricing Wheeled Power." National Regulatory Research Institute, Report NRRI-87–7.

McCabe, K., S. Rassenti, and V. Smith. 1991. "Experimental Research on Deregulation in Natural Gas Pipeline and Electricity Power Transmission Networks." In R.O. Zerbe and V.P. Goldberg, eds., *Research in Law and Economics*. Greenwich, CT: JAI Press.

Read, E.G., and D.P. Sell. 1988. "Pricing and Operation of Transmission Services: Short Run Aspects" and "A Framework for Transmission Pricing." Reports to Trans Power, Canterbury University and Arthur Young, New Zealand.

Read, E.G. 1988. "Pricing of Transmission Services: Long Run Aspects." Report to Trans Power, Canterbury University, New Zealand.

Smith, V. 1988. "Electric Power Deregulation: Background and Prospects." *Contemporary Policy Issues*, 6(3): 14–24.

Schweppe, F., M. Caramanis, R. Tabors, and R. Bohn. 1988. *Spot Pricing of Electricity*. Norwell, MA: Kluwer.

Williamson, O. 1966. "Peak-Load Pricing and Optimal Capacity Under Indivisibility Constraints." *American Economic Review*, 56: 810–827.

12 REGIONAL TRANSMISSION GROUPS: HOW SHOULD THEY BE STRUCTURED AND WHAT SHOULD THEY DO?

Edward Kahn

12.1. Introduction

Electricity transmission raises numerous challenges of combining cooperation and competition in a coherent institutional structure. It is widely recognized that existing forms of cooperative activity are not sufficiently tuned
to the rising competitive forces in the wholesale electricity market. These
competitive forces motivated the transmission access provisions of the 1992
Energy Policy Act. Sorting out the details of the new access regime could
entail costly and extensive litigation. As an alternative to litigation, some
form of voluntary negotiated activity among industry participants may be
a more efficient approach. Those interested in pursuing such a negotiation
framework have focused on a notion known as a *regional transmission group*
(RTG) as a new way of addressing the issues of common interest in
electricity transmission in a regime of broader access. While there is widespread perception that new forms of cooperation are desirable, there is
much less consensus on the detailed structure that such institutions should
take. This chapter examines these questions with particular emphasis on
the transmission capacity expansion problem. While this problem would
not be the exclusive focus of an RTG, it raises questions that have not

been well handled even by reform proposals addressing access issues involving only existing assets.

The discussion is structured in the following fashion. The history of
regulatory action in the U.S. electricity transmission market is reviewed in
Section 12.2 to explain the origin of the RTG concept. Section 12.3 discusses the procedural issues associated with establishing an RTG. These
include the role of state and federal regulation, the appropriate size of an
RTG, governance (voting rules and dispute resolution), and the relationship of RTGs with other cooperative organizations (regional reliability
councils and power pools). Procedural issues are important because they
illustrate both the need for cooperation and some of the barriers to it.
Section 12.4 discusses operational issues facing RTGs. Section 12.5 addresses the long-term planning and capacity expansion problem, including
cost allocation for lumpy investments. Conclusions are offered in Section
12.6.

12.2. Background

Competition in the wholesale electricity market was introduced in the
United States with the Public Utilities Regulatory Policy Act (PURPA) of
1978. PURPA created a class of essentially unregulated wholesale suppliers
that were entitled to sell power to local franchised utilities under the principle of *avoided-cost pricing*. The private power sector developed quite
unevenly over the next decade. In some states, local policies led to uncon-
trolled growth; in others, there was virtually no activity at all. The most
balanced development occurred where utilities began to organize the market
around a competitive bidding structure for long-term power purchase
contracts. These developments are surveyed in Kahn and Gilbert (1993).

A major impediment to the development of competitive wholesale
markets for long-term power was the limitation on transmission access.
Regulatory authority to order transmission services for third-party entities,
known as *wheeling*, lies with the Federal Energy Regulatory Commission
(FERC). PURPA restricted FERC's ability to order wheeling for private
producers. Few utilities offered it on a voluntary basis. Therefore, private
producers were still basically captive to the local franchised utility as a
monopsony buyer. This limited project size, and the alternatives available
to buyers.

FERC was not indifferent to the development of competitive forces. A
number of initiatives broadened the range and scope of market-based
pricing and increased transmission access (Einhorn, 1990; Tenenbaum and

Henderson, 1991). In both merger cases and wholesale marketing activities, FERC conditioned its approval on an open-access policy toward wheeling. These case law initiatives did not amount to a comprehensive approach to transmission access policy. Such an approach requires a pricing regime and access rules that recognize both financial and opportunity costs and that address the technical limitations on the operation of the bulk power network.

The challenges of formulating a coherent policy were accelerated by the Energy Policy Act (EPA) of 1992. EPA gave FERC the explicit authority to order wheeling and required transmission owners both to file tariffs for transmission services (wheeling) and to disclose the relevant data about capacity limits that would inform an open-access regime. These new regulatory powers will be difficult for FERC to wield effectively because electricity transmission raises technical complexities substantially beyond the level that administrative agencies typically engage. Transmission owners may use their knowledge to disadvantage potential wheeling customers. Transmission access and pricing disputes may lead to complex litigation. Some indication of these complexities will be outlined in this chapter.

In place of regulation based on litigation of a technically complex evidentiary record, many potential participants in the transmission market hoped that voluntary cooperative mechanisms mights solve some of the problems posed by open access. This hope was the basis of the RTG concept. Some efforts were made to incorporate language on RTGs into the EPA, but this was unsuccessful. Instead, FERC issued a request for public comments on what became known as the consensus proposal to keep open the option for using RTGs as a supplement to litigation (Sutley, 1993). RTGs might help to define available capacity for wheeling, provide a dispute resolution forum, and give guidance to FERC on regional aspects of pricing.

Forming an RTG presents substantial institutional challenges. These are outlined in the following section.

12.3. Institution Formation: Procedural Aspects of Forming an RTG

12.3.1. The Role of Regulators in Regional Transmission Groups

Traditional voluntary and informal joint planning has characterized utility co-ordination in transmission since the 1960s. These arrangements have worked quite well within the constraints of the mission they were set up to achieve (Gilbert, Kahn, and White, 1993). While there is no definitive mission statement characterizing this activity, it can be summarized as being an effort to maximize the economic benefits of coordination among vertically integrated utilities within a framework of high system reliability. This system achieved these goals, but at the cost of excluding transmission have-nots from the club of participants. With the requirements of the Energy Policy Act, in particular the obligation to provide third-party transmission services, such exclusionary arrangements are no longer adequate.

Achieving cooperation among parties who are also competing is the new challenge facing voluntary informal associations. It may prove necessary to invoke regulatory constraints of one kind or another to achieve nondiscriminatory access, for both private power producers and transmission-dependent (typically publicly owned) utilities. The basic problem is that private producers compete directly with transmission-owning vertically integrated utilities for market share in electricity generation. Transmission-dependent utilities compete for end-use demand and have historically been wholesale customers of the vertically integrated private firms. The potential for cost-shifting by the transmission owner onto its competitors may be difficult to control on a purely voluntary basis.

Although transmission reforms will be largely regulated at the federal level, there will be important tasks for state regulators. Traditionally, state regulation governed the investment decisions of vertically integrated investor-owned (IOU) firms. The federal domain was restricted to wholesale transactions over the existing network. EPA will expand the domain of federal regulation into transmission planning. Neither state nor federal regulators have shown much interest in transmission planning issues until quite recently (Baldick and Kahn, 1992). Most state regulatory interest in investment has focused on generation decisions and more recently on demand-side management. The growth in demand for transmission access will inevitably mean that capacity expansion will be required, and state regulators will have to approve most such investments, since IOUs are the dominant owners of capacity and right of way. Because transmission necessarily involves more than a single utility, state regulators will have to become involved in the joint planning process that has traditionally been conducted on a private voluntary basis by transmission-owning utilities.

One model for such involvement is a formal mandated process directed explicitly by a regulatory agency. This model is best illustrated by the Wisconsin Advance Plan (WAP). The Public Service Commission of Wisconsin (PSCW) has statutory authority over both investor-owned and publicly owned utilities in the state (a relatively unusual situation). By law, the PSCW requires utilities to participate in joint planning with active regulatory involvement. Recently, the PSCW has extended its interest to transmission planning, directing the study of various ways to optimize interconnection capacity in the state (PSCW, 1991a, 1991b). It is not clear how well the WAP will deal with private power access, since there has been relatively little activity of this kind in the state until quite recently.

It may be useful to explore some form of state regulatory involvement with RTGs that lies between the extremes of state planning and pure voluntarism. Ultimately state regulators must be comfortable with the activities of an RTG, since they must approve many plans that an RTG would formulate. The trick is to get their participation, concerns, and blessings without undue interference. This may be achievable, for example, by allowing regulatory representation on an ex officio basis but without voting rights. Such a proposal is part of the scheme proposed by an association of utilities in the Western United States (WATSCO, 1993). The value of ex officio, nonvoting representation is a little like the benefits of police patrols on the highways. It has the effect of inducing good behavior with only the implied threat of sanctions.

At the federal level, any RTG implementation of the requirements of the Energy Policy Act will require explicit approval. The FERC could adopt a generic approach to the RTG concept, through a rulemaking on this subject. Because views differ widely on this subject ("What They Said," 1993), it is unlikely that any prescriptive formulation will come from the federal level before there has been ample opportunity for regional experimentation.

12.3.2. Appropriate Size of Regional Transmission Groups

The reality of electrical flows may be an important determinant of the appropriate size of an RTG. It is reasonable to expect that parties impacted by transmission transactions should have a common forum in which to negotiate compensation and common interest issues. Figure 12.1, taken from Mistr (1992), shows that very large geographic areas can be affected by particular transactions. In terms of the regional reliability councils, the transaction shown here (from PSI to Eastern PJM) involves not only ECAR and MAAC, but also SERC and NPCC as well. If transactions such as the one shown in Figure 12.1 were frequent, it would argue for a very large geographic scope for an RTG. If, on the other hand, transactions were more localized, then a smaller area would be appropriate. In the United States there are three electrically separate regions: the western United

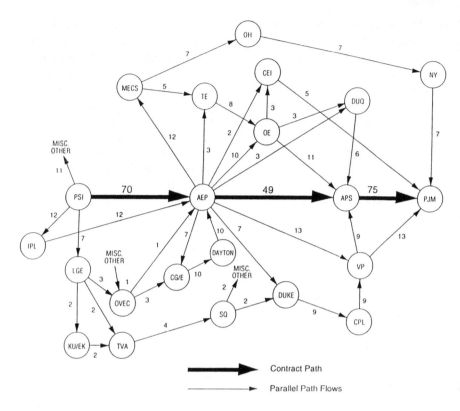

Figure 12.1. Transfer Response: 100 MW PSI to Eastern PJM (Based on a 1991 Summer Transmission Model)

States except for ERCOT, ERCOT itself, and the Eastern United States. In the limit then, this might define the smallest number of RTGs.

Procedurally, there is a tradeoff between a group that is too big to be managable and so small as to be either dominated by a single member, or irrelevant. In practice, the most serious problem is likely to come from a large number of entities. Suppose, for example, that three RTGs had members participating in a particular transaction. If each RTG used a different planning methodology and had different reliability standards, how would the transaction participants coordinate in constructing a single characterization of the effects of the proposed arrangement? Would these problems be any easier to resolve if all parties belonged to one RTG? While it might be easier to address the incompatability of analysis procedures within a single group, the probability of reaching consensus on any given issue would certainly decrease as the size of the group grew.

In the geographically largest NERC reliability region, WSCC, there are groups interested in RTGs in three subregions (the Northwest, the Southwest and California), Additionally, there are two different initiatives addressing regionwide issues (WSCC, 1993; WRTG, 1993). In this case at least the tradeoffs associated with different size organizations would not be complicated by incompatibile reliability standards.

Experience to date is quite varied. While no organization as yet would qualify as a fully functioning RTG (recognized as such by the FERC), there are groups of varying sizes that have been meeting with expectations of evolving in this direction.

12.3.3. Internal Organization of Regional Transmission Groups

Whatever the size of an RTG, and whatever form of regulatory participation and consultation that occurs, there must still be some form of governance. This amounts to specifying membership and voting rules. Membership should be open to participants with financial stakes in the process. Thus, in addition to public and investor-owned utilities, nonutility generators would be members of RTGs. As a practical matter, nonutility generators may frequently be represented by trade associations on a day-to-day basis, although such associations may not qualify as voting members.

The voting rule must lie somewhere between majority rule and a unanimity requirement. Cases where simple majority has been used can lead to substantial unhappiness on the part of minorities who may feel discriminated against. Cost-allocation disputes in MAPP have something of this quality, where small public systems feel that large transmission owners are charging excessively for system services (Bundy, 1993). At the opposite extreme, the New York power pool's unanimity rule has led to a number of cases where one member has blocked actions that would otherwise be beneficial in the large.[1] Voting rules can be formulated on a one member–one vote principle, use some kind of weighting system to represent the relative size of participants, or allocate votes to organization types.

Whatever the voting rule is, decisions must have some kind of binding effect on members. Even if RTGs do not have the legal authority to bind members, their procedures must help reduce transaction costs to be worthwhile. If a member disagrees with an RTG decision, there must be an arbitration process that is binding. One generally agreed upon goal for RTGs is a dispute resolution process that would limit litigation and regulatory adjudication. A practical issue in the specification of an arbitration procedure is the question of which issues should be resolved by regulatory processes and which should be settled by arbitration. Some of the current disputes about the WATSCO proposal center around these questions (Imparato and Ellison, personal communication, 1993).

RTGs will function to a certain extent as study groups where parties with different interests can meet to share information and seek joint economies. A major unknown is the extent to which these exchanges will become genuine forums for joint planning or simply more like information clearing houses. If the actual joint-study function is limited, then there may be more need for dispute resolution through arbitration or regulatory intervention. Simple information exchange might end up functioning as a form of prelitigation discovery. If RTGs are to serve a more cooperative role, they will need to have more than a passive information brokering function. The main reason for this is the technical difficulty of transmission planning. In a hostile and litigious atmosphere, it is easy to confuse technical difficulty with the exercise of market power. Therefore, some kind of analysis standards will have to be set for meaningful comparison of alternative proposals. The results of such comparisons need not be mandated presciptions, but at least the results of consistent and reasonably impartial study will be available for all participants, who will then make decisions with reasonably full information.

There are substantial barriers to achieving a real cooperative planning function. Some participants will have had more experience with such studies than others. Private producers and transmission-dependent utilities, in particular, will have had less experience of this kind than transmission owners. Yet they have the most to gain and lose from the success or failure of RTGs. Further, they probably have relatively larger incremental costs of participation than utilities. The principal cost associated with participation is developing the expertise to take a fully equal part in technical studies. This will involve acquiring and mastering engineering skills roughly equal to those of the transmission owning utilities. It can be expected that a certain amount of the bargaining and negotiating that will occur in RTG study groups will be in the form of manipulated technical assumptions and modeling techniques. Such manipulation is feasible since technical problems are never specified without some judgment and even ambiguity. This kind of strategic behavior has occurred in the avoided cost pricing arena, where the same kind of market-share conflicts arise between utilities and private producers (Kahn, 1993; Pechman, 1993).

12.3.4. Relationship to Reliability Councils

It may be a good idea to separate the existing set of regional reliability councils from any system of RTGs. Reliability is a constraint within which economic cooperation and competition must function. If the reliability councils took on the economic functions of an RTG, the competitive tensions might have unfortunate implications for both reliability and competition. Separating the economic and reliability functions, however, raises the risk that a coalition of firms might be able to use reliability arguments for strategic purposes. A plausible scenario embodying this kind of manipulation would be a strategy of cost shifting based on a bogus but difficult to disprove claim that incremental line loadings associated with a transmission service request would raise reliability problems. The purported problems would require expensive mitigation costs, and such costs would have to be borne by competitors of the transmission owners. Disproving such claims or proposing alternative remedies would be expensive and difficult. More positively, however, it might be difficult to sustain collusion of this kind in the long run, since there would be incentives for a disaffected participant to blow the whistle on such arrangements when they turned out to be to his own disadvantage.

Even though separation of the reliability and economic functions is probably best, there would still be a need for coordination and communication between RTGs and reliability councils. Such coordination would be important to the RTG's evaluation of capacity additions. Coordination might get particularly difficult in the case where an RTG was larger than a single reliability council area. Then there might be more than one reliability standard (or set of standards) to which the RTG would have to respond.

One regional council, the Western Systems Coordinating Council (WSCC) has proposed a regional planning function that would perform some of the same planning functions that an RTG might potentially perform. In its proposal (WSCC, 1993), there is language about the standardization and consistency of planning studies. But the proposal explicitly renounces economic evaluations, comparative selection of competing projects, or project-ranking activities. In the same vein, the WSCC proposal language about dispute resolution is oriented to mediation and advice, although it allows for binding arbitration. The WSCC would coordinate with RTG activity at the subregional level.

Ultimately, coordinated planning must confront the very activities that the WSCC proposal seeks to avoid, the economic comparison of alternative projects. The alternative to such comparisons is an implicit agreement to settle for a more expensive and inefficient network than necessary, on the grounds that the effect on total costs is small. While such arguments are occasionally made off the record, it is difficult to believe that state regulators, who must approve new transmission investment would willingly accept a process with inefficient outcomes as its design criterion. If, therefore, economic comparison of alternatives is ultimately necessary, the WSCC approach goes only half of the way. Given the potential for strategic use of technical data in conflicts over such evaluations, the WSCC approach could provide a technical referee function. While the refereeing function can significantly narrow the range of disputes and facilitate conflict resolution in a regulatory setting (Kahn, 1993), it is not clear whether the WSCC approach would achieve this.

There is an alternative approach to regional planning in the WSCC that would form a more active RTG (WRTG, 1993). This proposal is silent on the question of analysis standards for planning studies, but such an approach is not bounded out of the discussion. While such standards may have ambiguities and fail to capture all economic factors, they would at least ensure a reasonably complete comparison of alternatives. Such analysis would have no more than moral suasion in this proposal. If a coalition of participants was willing and able to proceed with what looked like an inferior project, there would be nothing to prevent them.

12.3.5. Power Pools as Regional Transmission Groups

Organized power pooling organizations are natural candidates to extend their coordinating activities to include the functions of an RTG. There is considerable variation in the extent of formal pooling activity in the United States (Palmero, 1991). Where pools are strong and perform central dispatch and generation planning coordination, the RTG functions could fit naturally into the existing structure of governance and administration. Grafting transmission access, planning, and pricing onto a pool structure would still involve resolution of the competitive conflicts among the parties. Would membership be extended to private producers in such a setting? Would preferential access conditions for original members be accomodated?

The one attempt to graft RTG functions onto an existing power pool structure was initiated in connection with the Northeast Utilities merger with Public Service of New Hampshire. The agreement which was negotiated among the New England utilities (Northeast Utilities, 1993) subsequently splintered due to disputes over pricing.

Where formal pooling institutions are weak, RTG activity will have no other institution on which to build.

12.4. Operational Issues

Pricing transmission services is a major feature of the FERC's responsibility in implementing EPA. One substantial challenge for pricing is reflecting the operating characteristics of the bulk power network. This could be a significant activity for RTGs because it will be important to develop a regional perspective on technical issues such as interface constraints, voltage and stability requirements, and parallel flows.

Economic efficiency requires achieving a consistency between operational characteristics of the bulk power network and the price signals sent to users of the system. There are many levels of approximation that can be used to achieve this consistency. One of the simpler steps is to use zoned rates for contract paths. This was proposed in the New England RTA (Northeast Utilities, 1993). Zoning still preserves the fiction of contract paths (that power actually flows along some designated path between seller and buyer) but allows for some recognition of interface constraints. The parallel flow problem, illustrated in Figure 12.1, is widely recognized as posing the greatest challenge to efficient pricing.

Engineering approaches to accounting for parallel flows have been suggested (Mistr, 1992; Kovacs and Leverett, in press). These methods would use load flow simulations to assign cost responsibility, primarily for pricing firm transmission service.[2] Such analysis could be appropriately performed by an RTG, since the parallel flow problem is regional. The results of such studies will be highly dependent on the base case specification of system operation. More important, the outcomes for particular transmission users will depend strongly on the order in which such studies are conducted. Broadly speaking, effects would be less severe for transactions modeled with a less loaded system than for a more heavily loaded system. Therefore, pricing using these one-at-a-time engineering calculations will raise issues about the exact order of transmission requests. Disputes about which requests precede which other requests will arise. The inefficiency of first-come, first-served allocation, familiar from PURPA experience, will appear in this domain. Uniform practices within an RTG may alleviate some of the controversies about the priority of specific requests, but not the efficiency issues.

The *node pricing* approach first proposed by Schweppe, Caramanis, Tabors, and Bohn (1988) and advocated more recently by Hogan (1992) can potentially solve the efficiency problems associated with the engineering approach to allocating costs in the face of parallel flows. Node pricing would set the transmission price between any two points in the network simply at the difference between marginal generation costs at those points. Assuming that the network is operated efficiently, this pricing approach correctly reflects the costs of all network constraints.

The pattern of network flows, however, is influenced strongly by the pattern of generation as well as the configuration of lines. This fact is not widely appreciated outside of the engineering community. The joint determination of power flows by both network configuration and generation pattern can create local geographic monopoly power. Situations of this kind can occur in any market structure. In Great Britain, the Office of Electricity Regulation has inquired into some such cases, where the bidding behavior of certain generators appeared to exploit local geographic monopoly power. When the National Grid Company attempts to implement the unconstrained least cost dispatch, it must sometimes dispatch out-of-merit-order plants to relieve transmission constraints. Under the pool pricing rules, such plants are paid their bid price, which is necessarily higher than short-run marginal cost (if it were not higher, the plant would have been dispatched in the unconstrained case). This phenomenon has led to substantial payments for constrained-on plants. Indeed, the bidding behavior of some of these plants is suspicious; involving, in some cases, a tripling of the bid price. Local area constraints can be relieved by a number of methods including network reinforcements or load management, which may be much cheaper than present practice (OFFER, 1992). Thus, operations and planning are not so easily separable as one might like.

Therefore, node pricing will only function properly if the system dispatch is efficient. In a centralized system, there is good reason to believe that dispatch will be efficient, especially if regulators can audit costs. This need not necessarily be the case in a decentralized world of many control centers. In this case, local monopoly power can distort the potential benefits of the node pricing approach. A simple example of this is given in Kahn and Gilbert (1993), where the transmission owner misrepresents his marginal cost as equal to a monopoly price, resulting in an inefficiently high node price for transmission. For the United States, we might expect locally efficient dispatch, but there can be some question about global efficiency beyond the single control area.[3] RTGs might usefully provide a forum in which operating procedures are discussed among the participants. Such discussions might lead to the identification of procedures that were more economic for all parties than for parties individually. If this turned out to be the case, some forms of compensation might be required

Table 12.1. Scale Economies and Lumpiness: SCE Data

Substation bus	Incremental capacity (MW)	1997 Incremental cost (10^3 $)	Cost per kW($) Maximum	Cost per kW($) Minimum
Coachella 230 kV	0–25	0	0	0
	26–565	$48,100	$1,850	$85
	566–750	97,900	173	131
Kramer 230 kV	0–35	0	0	0
	36–245	47,300	1,314	193
	246–750	69,700	283	93
Antelope 230 kV	0–150	0	0	0
	151–500	22,800	151	46

Table 12.2. Economies of Scale by Voltage Class

Voltage class	Capacity (MVA)	Average length (miles)	Cost per mile ($)	Cost per MW-mile ($)
500 kV	2600	50	$1,151,800	$443
230 kV	600	18	805,800	1,343
115/138 kV	256	13	556,756	2,175
34.5 kV	36	20	75,717	2,103
13.2/12.5 kV	13	8	75,717	5,824

to implement changes. Alternatively, different perspectives on operating practice could escalate into disputes.

12.5. Planning Issues: Capacity Expansion, Scale Economies, and Cost Responsibility

12.5.1. Technology Fundamentals: Lumpiness and Scale Economies

Generically, transmission technology exhibits both scale economies and lumpiness. These features are illustrated in Tables 12.1 and 12.2. Table 12.1 is a selection from the "telephone book" produced by Southern California Edison (SCE) for use in California's private power solicitation process (SCE, 1992). The purpose of this document is to convey to potential bidders the cost (if any) of upgrades to the SCE high-voltage network required to accommodate incremental generation at various substation buses. These costs will be used in bid evaluation. Table 12.1 gives these costs for three locations. The complete list covers seventy-two locations. Only twenty of them require capacity expansion for less than 500 MW additional generation.[4]

The data in Table 12.1 shows costs within a single voltage class. The capacity expansion within a voltage class would typically be achieved by adding circuits to existing towers where possible, expanding a given corridor with parallel towers, or adding reactive power equipment to relieve a voltage constraint. In all cases, the capacity increments are lumpy rather than continuous. Therefore the costs per unit of capacity for a given increment vary from a maximum to a minimum level depending on the utilization. This has implications for cost allocation, which are discussed below. Finally, unit costs may either increase or decline as capacity is expanded.

Table 12.2 shows the scale economy across voltage classes. This data, taken from Mistr (1992) is reasonably representative of other estimates (for example, Baldick and Kahn, 1992). The cost/mile is easily estimated. The important source of variation in cost per MW-mile estimates is the capacity rating assigned to a particular line.

The ratings in Table 12.2 are typically thermal limits. This is the least binding constraint. Quite frequently capacity limits are due to voltage or stability problems; these constraints depend on the configuration of the network, including the pattern of loads and generation. It is the difficulty of characterizing the capacity limits of lines subject to reliability constraints that is responsible for many of the policy problems in transmission. Where long lines are involved (200 to 300 miles), for example, capacity may be about half the value listed in Table 12.2, and therefore cost per MW-mile would be twice the value given.[5] Even in such cases, there are still scale economies by voltage class.

12.5.2. Cost Allocation

In light of the indivisibilities illustrated by Tables 12.1 and 12.2, there is a considerable potential for a mismatch between the demand for incremental transmission capacity and the supply. Given the obligation of transmission owners to provide capacity, such mismatches imply that there will be excess capacity. This situation raises questions of cost allocation and risk that RTGs will have to address.

Within the service territory of a single large utility these issues have already arisen. One example involves Southern California Edison. SCE prepared the analysis (excerpted in Table 12.1) for a private power solicitation. Their position on the cost allocation issue is that the entire cost of a transmission upgrade would be the responsibility of a bidder whose project required any part of the incremental capacity. SCE has argued that such an allocation in the face of uncertainty about future demand for such capacity is efficient (Hunt, 1992). What is not clear under such proposals is whether the private producer would then have resale rights for the capacity in excess of his demands.

The same cost allocation and resale rights issue would arise in an RTG context. There might be some better smoothing of supply and demand in an RTG, since more potential users would become aware of the availability of capacity and so could participate at the outset. Assuming that mismatches persist nonetheless, the efficient solution would be to allocate secondary market rights to those users who must bear the cost responsibility for lumpy incremental capacity additions.

There is a rough analogy here with the natural gas pipeline industry, where a similar problem exists. A mismatch between holders of pipeline capacity and potential users of that capacity arose as the gas industry experienced restructuring during the 1980s. It was argued that some kind of market process should be used to match capacity and demand better (Alger and Toman, 1990). The FERC did move in that direction with Order No. 636 (FERC, 1992); although it is not yet clear how successful such programs will turn out to be. This situation differs from electricity in that the capacity allocation problem in pipelines refers to existing assets rather than incremental capacity expansion, but the same fundamental problems of efficient use of property rights are similar to both settings.

An alternative approach to cost allocation lies in focusing RTG activity on a long-range planning process. Utilities would forecast their need for transmission capacity, and plan accordingly on a joint basis. Any new demand for transmission service would then amount to a modification of the RTG plan. Most such demands would involve accelerating the service date of planned additions. The compensation to utilities from such users would be limited to the cost of advancing the construction of facilities that would have been built later. This kind of cost allocation is essentially an "interest only" charge to the transmission service user. It amounts to a mechanism for allocating a pro rata share of indivisible facilities. In this kind of framework, the cost burden of capacity expansion would be reduced for new users, and the responsibility of utility ratepayers would increase compared to the first-user-pays-all option. A mechanism of this kind amounts to a risk sharing between ratepayers of the transmission owning utility and other users. Such sharing should encourage utilities involved in the forecasting of future needs to be as attentive and accurate in their planning as possible.

A number of difficult issues would arise in an interest-only approach to cost allocation. First is the believability of the transmission planning process. Planning for generation capacity has proved to be a highly error-prone process for electric utilities in the past. The demands of a transmission planning process are even greater, and so consequently is the room for error. Given the competitive conflicts among participants, the potential for strategic manipulation in the planning exercise cannot be ignored either. This issue involves not the accuracy of the plan but the level of commitment made to the plan by its sponsors. The WSCC planning process discussed in Section 12.3 would not have the level of commitment necessary to implement an interest-only approach to cost allocation because the participants would not be seeking the best plan but only exchanging information in an essentially informal process.

Planning in an RTG context has some similarities to the method proposed by Frame (1992) in his contract approach to transmission service pricing. Both are based on complex planning studies, relying on a change-in-plan concept for determining cost. The RTG setting could substantially reduce the third-party and regulatory concerns of arbitrary or excessive costs for capacity expansion identified by Frame, since there would be a lot of focused attention on the planning process by RTG members. The opportunity cost concepts advocated by Frame, however, would be unlikely to be accepted by transmisssion service requestors.

Assuming that the planning and forecasting process worked tolerably, there is still a problem of residual claims on excess capacity. In the interest-only framework, the transmission-building utility would own and pay for the excess capacity. In subsequent expansions of the generation mix, it becomes questionable whether this capacity would be reserved for the utiliy's future use or could be deemed available for nonutility suppliers. This future use problem makes clear the underlying uncertainty about the use and purpose of the transmission network in the United States. This uncertainty is fundamentally due to the ambiguities surrounding the market structure in generation and the future role of vertical integration. These issues are addressed briefly in the following section.

12.6. Conclusions

Growing competition in the wholesale electricity markets is raising awareness of the need for more coordination among participants in the transmission segment. RTGs are one response to this pressure but not as yet a well-defined response. The discussion of procedural issues involved in the formation of an RTG illustrates the wide range of stakeholders interested in regional transmission issues and the barriers to cooperation. The alternative to cooperation is costly litigation. It may be inevitable that the costs of litigation will have to become apparent before the benefits of cooperation become clear to all parties. The discussion of potential operational and planning benefits that can accrue from cooperation shows that the effort to form new institutions should be worthwhile. They key question for the RTG concept is whether cooperative activity can arise on an essentially voluntary basis. If the need for cooperation cannot be supplied without government intervention, then some sort of mandated activity or structural reform will be supplied by the political process.

This chapter has not answered the questions addressed in its title. Definitive answers may only emerge from industry practice. Alternatively, such answers may require a further structural realignment in the electricity industry, perhaps along the lines of the British model. In this case, what we might call the strong version of an RTG, perhaps involving spinoffs or even some government ownership, would become indistinguishable from the functions of the National Grid Company. The viability of a less radical outcome will depend on the creativity of the traditional muddling-through process of American political economy.

Acknowledgments

This research was funded by the Assistant Secretary for Energy Efficiency and Renewable Energy, Office of Utility Technologies, Office of Energy Management of the U.S. Department of Energy under Contract No. DE-AC03–76SF0098 and by the Universitywide Energy Research Group of the University of California. I have benefitted from discussions with Richard Gilbert, University of California–Berkeley; Steve Henderson and Margaret Jess, Federal Energy Regulatory Commission; Lon Peters, Public Power Council; Alfred Mistr, Virginia Power; Bill Stillinger, Northeast Utilities; Jeanine Hull, LG&E Power Systems; Carl Imparato, Pacific Gas and Electric Company; and Chris Ellison, Independent Energy Producers.

Notes

1. One example of this, cited in the NYPSC (1991) audit of the New York power pool, involved the operation of a jointly owned pumped storage facility. Each member has its

pro-rata share of scheduling rights on the facility. The usage pattern under this decentralized scheduling was found to be less efficient than if centralized operation were used. A centralized scheme, however, would have reduced the benefits to one participant and hence was blocked under the unanimity rule.

2. For short-term transactions, or for cases where holders of transmission rights do not need to use their entitlements, some kind of bidding system could be used to reallocate capacity. This is similar to capacity release mechanisms in natural gas pipeline regulation.

3. An interesting simulation study illustrating this kind of effect focuses on coordination economies between control areas involving unit commitment (the operator's decision to turn generating units on and off) (Lee and Feng, 1992).

4. Table 12.1 does not reflect the final form of this data for use in bidding. That is given in SCE (1993). This data shows the effect of lumpiness more clearly than the final version. It is also interesting to note that the upgrade cost estimates have gone through numerous iterations and changes, reflecting either the difficulty of making such estimates or the amenability of the process to strategic manipulation, or both.

5. Line capacity as a function of line length is discussed in Dunlop, Gutman, and Marchenko (1979).

References

Alger, D., and M. Toman. 1990. "Market-Based Regulation of Natural Gas Pipelines." *Journal of Regulatory Economics*, 2(3): 263–280.

Baldick, R., and E. Kahn. 1992. "Transmission Planning in the Era of Integrated Resource Planning: A Survey of Recent Cases." LBL-32231.

———. 1993. "Network Costs and the Regulation of Wholesale Competition in Electricity." *Journal of Regulatory Economics*, 5(4): 367–384.

Bundy, T. 1993. "MAPP MW-Mile Method: 'The Devils Are in the Details.' " Paper presented at the TAPS Conference on Distance Based Transmission Pricing Issues.

Dunlop, R., R. Gutman, and P. Marchenko. 1979. "Analytical Development of Loadability Characteristics for EHV and UHV Transmission Lines." *IEEE Transactions on Power Apparatus and Systems*, PAS-98: 606–617.

Einhorn, M. 1990. "Electricity Wheeling and Incentive Regulation." *Journal of Regulatory Economics*, 2(2): 173–189.

"What They Said About Regional Transmission Groups." 1993. *Electricity Journal*, 6(2): 30–39.

Federal Energy Regulatory Commission (FERC). 1992. Order No. 636.

Frame, R. 1992. "Transmission Access and Pricing: What Does a Good 'Open Access' System Look Like?" NERA Working Paper.

Gilbert, R., E. Kahn, and M. White. 1993. "The Efficiency of Market Coordination: Evidence from Wholesale Electric Power Pools." University of California Energy Institute Working Paper, PWP-012.

Hogan, W. 1992. "Contract Networks for Electric Power Transmission." *Journal of Regulatory Economics*, 4(3): 211–242.

Hunt, P. 1992. "Supply Procurement for Uncertain Future Demand Under Regulation." Rutgers University Advanced Workshop on Regulation and Public Utility Economics.

Kahn, E. 1993. "Regulation by Simulation: The Role of Production Cost Models in Electricity Planning and Pricing." University of California Energy Institute Working Paper, PWP-014.

Kahn, E., and R. Gilbert. 1993. Competition and Institutional Change in U.S. Electric Power Regulation, University of California Energy Institute Working Paper, PWP-011.

Kovacs, R., and A. Leverett. In press. "A Load Flow Based Method for Calculating Embedded, Incremental and Marginal Cost of Transmission Capacity." *IEEE Transactions on Power Systems.*

Lee, F., and Q. Feng. 1992. "Multi-Area Unit Commitment." *IEEE Transactions on Power Systems*, 7(2): 591–599.

Mistr, A. 1992. "A Proposal for Fundamental Reform of Transmission Pricing." Virginia Power.

New York Public Service Commission (NYPSC). 1991. "Report on the Management and Operations Audit of the New York Power Pool."

Northeast Utilities. 1993. "Agreement with Respect to Regional Transmission Arrangement for the New England Power Pool."

Office of Electricity Regulation (OFFER). 1992. "Report on Constrained-On Plant."

Palmero, P. 1991. "Institutional Issues Regarding Centralized Dispatch." Appendix A in Public Service Commission of Wisconsin (PSCW), Advance Plan 6 Centralized Dispatch Study: Report (D26).

Pechman, C. 1993. *Regulating Power: The Economics of Electricity in the Information Age.* Boston: Kluwer.

Public Service Commission of Wisconsin (PSCW). 1991a. "Advance Plan 6 Interface Study: Transmission Report (D23s)."

————. 1991b. "Advance Plan 6 Interface Study: Resource Related Benefits of Increased Interface Transmission Capability (D22c)."

Schweppe, F., M. Caramanis, R. Tabors, and R. Bohn. 1988. *Spot Pricing of Electricity.* Boston: Kluwer.

Southern California Edison (SCE). 1992. "Draft Transmission Cost Tables." Revised October 1.

————. 1993. "Transmission Cost Tables." August 11.

Sutley, N. 1993. "Capacity Planning and Regional Transmission Groups." Paper presented at the Wheeling and Dealing Conference.

Tenenbaum, B., and S. Henderson. 1991. "Market-Based Pricing of Wholesale Electric Services." *Electricity Journal*, 4(10): 30–45.

Western Association for Transmission Systems Coordination (WATSCO). 1993. "Draft Bylaws."

Western Systems Coordinating Council (WSCC). 1993. "A Process for Regional Planning in the WSCC Region." WSCC Regional Planning Policy Committee.

Western Regional Transmission Group (WRTG). 1993. "Draft Bylaws." September.

13 PROFIT SHARING REGULATION OF ELECTRICAL TRANSMISSION AND DISTRIBUTION COMPANIES

by Ingo Vogelsang

13.1. Introduction

While competition appears to be feasible and efficient for electricity generation, distribution and, to some extent, transmission maintain properties of a strong natural monopoly and are not considered to be contestable in the sense of Baumol, Panzar, and Willig (1982). As a consequence, distribution should be and transmission may have to be regulated by legal entry barriers and through limitations of price-setting behavior of the firm. The question is what is the best method of price regulation in these areas.

13.1.1. Distribution

Distribution companies are highly capital intensive and employ long-lived assets. For the last two decades, their growth has been slow in the United States and Western Europe, implying that a major part of their capital stock is old and has low historic costs. This does not hold, though, for other parts of the world, such as the Pacific rim countries. In addition, investment for new distribution capacity can be quite lumpy. Like in many network industries there are economies of density, meaning that, ceteris paribus, average costs per customer decline as customer density grows.

By far the main variable input of distribution companies is electricity that they can buy from the transmission network or directly from generating companies, with transmission as a separate item. If electricity is purchased by a distribution company under long-term agreements with generation and transmission companies, its characterization as a variable input is called into question, and this can affect the best regulatory setup. Other variable inputs include labor and materials.

Electricity distribution companies are multiproduct firms providing electricity at different voltage levels, at different locations and times, providing distribution bundled with electricity or as a separate item. Worldwide, many distribution companies are vertically integrated backward into transmission and generation. In this chapter, however, we consider only distribution companies as legally separate and under different ownership from transmission and generation.[1]

13.1.2. Transmission

The electricity transmission lines connect generation facilities with distribution networks. While single transmission lines could, in principle, connect each generating plant with a distribution network, transmission usually forms itself an interconnected network. This way each generating plant can interconnect with each distribution network, and choice becomes available on both ends. A distribution network can have choice among different suppliers of generation, and a generating firm can sell to several distribution networks. It is the economies of scope through interconnection and the parallel running of high-voltage transmission lines that tend to give the transmission network properties of a natural monopoly in the sense that duplication would lead to higher costs (see, for example, Joskow and Schmalensee, 1983). In addition, the wheeling of electricity through a transmission network is a complex commodity in that the network continuously has to be kept in a state of equilibrium. Nevertheless, it is quite conceivable that an interconnected transmission network could consist of several legally and commercially independent but interconnected parts. In particular, the transmission line interconnecting a generating plant with the rest of the transmission network could well be vertically integrated with that generating plant and vertically separated from the rest of the transmission network. Separation of a transmission network into several subnetworks could lead to some choice between transmission suppliers when it comes to wheeling of electricity over larger distances. Thus, the argument against competitive and in favor of regulated monopoly supply is somewhat weaker in the case of transmission than it is in the case of distribution networks.

Vertical integration between transmission network and distribution networks appears to be nonoptimal, assuming that the latter are so localized that a large number of them could be served by a single integrated transmission network. Also, the transmission grid is the key to development of a competitive market for power. The transmission system provides the bottleneck access of a generating facility to distribution networks and industrial customers. Under vertical integration between transmission and distribution competitive generators would face a monopsony, and thus the power market would not become competitive.[2]

What makes regulation of transmission networks particularly challenging is attempting to set incentives in such a way that the transmission network ideally complements generation and distribution. This includes minimizing distances between power stations and demand centers for competitive alternatives, providing system reliability (frequency and voltage levels), smoothing load patterns, coordinating maintenance of power plants, and providing emergency responses (Joskow and Schmalensee, 1983; Joskow, 1993). All this has to be achieved for a commodity that is exceedingly hard to cost out.

13.1.3. Regulatory Objectives and Their Implementation

What are the objectives of regulating electricity distribution and transmission companies? Each company has to be induced to build and maintain a network that reliably provides electricity at low costs at the points demanded. Thus, the utility has to be provided with incentives for production and procurement efficiency. Production efficiency relates particularly to the building and maintenance of the network. Procurement efficiency relates particularly to a distribution company's purchase of electric power and transmission services. In addition, pricing needs to be efficient and fair (equitable). Efficiency in pricing means that consumers of the network services be induced to make the right complementary investments and consume allocatively efficient quantities. For transmission networks these complementary investments relate to generation facilities and distribution networks. Equity and fairness include a variety of postulates, such as the availability of lifeline rates and prices that do not exceed consumers' stand-alone costs. Since the transmission network only provides intermediate inputs to distribution networks and large industrial consumers, the fairness postulate for transmission rates applies predominantly to nondiscrimination between equally situated buyers (who may be competing with each other).

How can one design a regulatory system that best fulfills these regulatory objectives? Designing regulation involves basic and detail regulatory engineering (Levy and Spiller, 1993). Basic regulatory engineering refers to the creation of a regulatory body that can execute regulatory rules designed in detail engineering. Basic engineering thus helps decide whether the regulatory body should be part of government or independent or whether regulatory decisions are to be based on a contract between the utility and government (in the form of a license) or simply on a regulatory law. Detail engineering refers to specific regulatory methods, including the use of price-cap versus rate-of-return regulation versus profit-sharing regulation. Both, the choices in basic and in detail engineering depend on the type of industry to be regulated and on the institutional traditions and capabilities of a country, in particular, on its commitment capabilities.

Commitment is particularly important in the electricity sector because most of the capital equipment installed by electric utilities is sunk. After investment has occurred, regulators may want to expropriate utility investors by not granting rate increases that would cover capital costs. Although this may be in the short-run interest of electricity consumers, it is against their long-run interest since it provides disincentives for further investments. Thus, before investing, the electric utility will want regulatory commitment against such expropriation. At the same time consumers want to avoid regulatory capture by the regulated firms and both, consumers and utilities want flexibility in adapting to changing conditions, such as fluctuating demand or technical progress. Thus, regulatory discretion needs to be contained in order to provide commitment and to avoid regulatory capture. At the same time we want regulatory flexibility. As shown by Levy and Spiller (1993) the fulfilment of these postulates may be infeasible in a country with insufficient commitment and administrative capabilities. An example for the lack of both capabilities, discussed in Levy and Spiller, is the telecommunications sector in the Philippines, where regulation has been highly unstable, resulting in substantial underinvestment. This is surprising insofar as telecommunications is an expanding sector and the Philippines have not been stagnating economically. The country has, however, fallen substantially behind other countries in the region.

If only the administrative capabilities are missing in a country the fulfilment of the postulates may be possible but come at the expense of efficiency or equity. An example for this is the telecommunications sector in Jamaica that is regulated under a 25-year monopoly license with a very crude rate-of-return constraint. This virtually guarantees the incumbent monopolists substantial returns and has led to high rates of investment.

What are the specific commitment issues in electricity? The most important commitment problems in recent history have arisen with respect to generation capacity, especially for nuclear power but also for coal- and oil-fired stations. There are a number of factors that have contributed to these problems. They include environmental and licensing problems that have led to increasing lead times and vast cost overruns. They include exaggerated demand projections, fuel switching away from oil, and problems of operating performance with new technologies and large scales. In several countries these problems have resulted in the emergence of small-scale power producers and the reemergence of industrial cogeneration. Smaller scales reduce lead times and lumpiness and allow for competition in electricity generation. While this does not necessarily eliminate regulatory commitment problems, the emergence of a larger set of actors may hold regulatory expropriation at bay.

Transmission capacity follows investment in generation capacity and thus shares some of the commitment problems of the generation stage. Transmission, however, requires less lumpy investment and shorter lead times and is more flexibly employed. Not all new transmission capacity is sunk if a new power station becomes a stranded investment. Also, the environmental and technological uncertainties surrounding transmission appear to be less serious.

Distribution, at first glance, appears to involve less regulatory commitment problems than the other two stages in electricity. The investment in distribution capacity is mostly demand driven; in the last few decades technical progress has not been a major source of uncertainty; cost uncertainty is mainly in underground construction, but not exceedingly severe. However, there is a potentially very serious commitment problem surrounding long-term supply contracts in electricity. These commitment problems have to do with length and type of supply contracts. For example, a cost-plus contract with a nuclear power plant yet to be built could generate commitment problems quite similar to the ones faced by the generating company. The U.S. experience with long-term contracts for natural gas by pipeline companies may serve as a prototype for commitment problems raised by supply contracts. These contracts contained take-or-pay clauses that ordinarily would have forced the pipeline companies to take delivery of the gas they had contracted for. Declining demand, however, made them attempt to renege on these contracts with the regulators as mediators. Regulators, with the benefit of hindsight, are likely to challenge long-term contracts that impose high costs or surplus quantities just as if the distribution company had (in the eyes of the regulator) wrongly invested in generation and transmission capacity. On the other hand, lack of long-term contracts can lead to supply shortages.

In this chapter we are not so much concerned with basic regulatory engineering but rather with the detail engineering of designing the appropriate pricing rule to be used by the regulator. We do not discuss an optimal regulatory approach suitable for every possible country but assume that, in a fictitious country, there exists a regulatory body that, under normal circumstances, is credible enough to implement the regulatory rule we choose but that does not have full commitment power under extreme circumstances, such as in cases of financial strangulation or exorbitant profits of the regulated firm. Thus, the regulator can commit neither to leaving high profits to the firm in case of high efficiency nor to letting the firm go bankrupt in case of low efficiency. This has an immediate effect on the power of feasible regulatory incentive schemes. High-powered incentive schemes may not be feasible because the regulator cannot commit to them in case of extreme outcomes.[3]

13.2. Price-Level Regulation

13.2.1. Introduction

It is convenient to address the regulation of price level and of price structure as two separate concerns. Regulating the firm's price level has the broad effect of distributing social surplus between the firm (profit) and its customers (consumer surplus). Absent competition or tax or subsidy schemes, it is the best tool to provide the firm with incentives to produce efficiently (adverse selection and moral hazard). It also determines the average allocative efficiency of the firm. In contrast, the firm's price structure distributes consumer surplus between various consumer groups. It also has a major effect on allocative efficiency and more than a second-order effect on the firm's profits.

We start by discussing mechanisms for price-level regulation. Here we differentiate between four idealized types of regulating a monopoly's price level: rate-or-return regulation, price-cap regulation, profit sharing, and yardstick regulation.

13.2.2. Rate-of-Return Regulation

Under rate-of-return regulation the regulator aims for a price level such that the firm's achieved rate of return remains at or below an allowed rate of return. Rate-of-return regulation has been criticized in the past mostly for the alleged lack of incentive for cost minimization, due to its cost-plus nature. On the positive side, rate-of-return regulation in the United States has evolved from a fairness doctrine that has provided it with substantial commitment power. This has not prevented slow adaptions to changing environments, something that came close to regulatory expropriation (Joskow, 1974; Joskow and MacAvoy, 1975). In the last few years this commitment power has, in the practice of electricity regulation, been challenged by applications of the used-and-useful criterion to public utility investments. Regulators have questioned and successfully denied cost overruns or entire investments in the rate base. While an efficiency rationale can be given to this type of regulatory behavior (Gilbert and Newbery, 1988; Lyon, 1991), it does raise moral-hazard issues and increases regulatory uncertainty. As a result, there may be underinvestment rather than overinvestment under rate-of-return regulation. This problem holds true in particular for investments with high-cost risks to begin with (such as nuclear power plants). Some investments in transmission lines may also be jeopardized as a consequence. Underinvestment appears to be less of a problem for electricity distribution networks. Here, the more traditional concern with overinvestment holds—that is, how to induce cost reductions and optimal capacity planning. It is not clear that the used-and-useful criterion would have much to contribute with respect to the distribution network. The cost-minimization issues in distribution networks are rather too subtle to be handled by outside cost controllers. Instead, the cost-plus nature of rate-of-return regulation is likely to make rate-of-return regulation of electricity distribution unattractive from an incentive perspective. Also, the intricacies of electricity supply arrangements for distribution companies make the firm's rate base the wrong lever to set incentives.

Although measuring a firm's cost of capital and the value of the rate base may be quite difficult, rate-of-return regulation is basically informationally simple and does not depend on information of regulators that is not, in principle, publicly available. The price for this simplicity and for high regulatory commitment is a lack of incentives.

13.2.3. Price Caps

Under price-cap regulation the firm's price level has to remain at or below a cap that moves over time, according to a prespecified formula. This formula usually contains three distinct elements:

- *A general adjustment factor for the economy's price level.* This inflation adjustment can be seen as representing the firm's unspecified
  input prices or, more likely, the inflationary loss of consumers. Since
  transmission and distribution companies have few cost items that could
  be linked to the current level of inflation, there is likely to be a major
  discrepancy between consumer interest in inflation-adjusted prices
  and the interest of such a company in cost-related prices. As a result
  a large fraction of the company's costs will have to be kept outside
  the inflation adjustment.
- *One or several adjustment factors for specific inputs or cost items that
  are passed through to consumers.* These include tax and fuel adjustments. For distribution companies the major question is, "To what
  extent should generating and transmission costs be passed through?"
  For both transmission and distribution companies, "Should there be
  an adjustment for economywide interest-rate changes?"
- *A general productivity adjustment factor, X.* Although, as a result of
  computerization and better load management, some productivity
  improvements of transmission and distribution companies can be expected over time, it is hard to find empirical evidence in favor of an
  X different from zero. In case of transmission companies this could
  change once wheeling over longer distances becomes more widespread.

Although the price-cap formula adjusts for cost and demand changes
that a regulated firm may experience, profits over time are likely to reach
positive or negative extremes that are unacceptable to the regulator. Since
such a situation has to be expected, price-cap formulas are usually revised
every few years. These revisions tend to be based on the firm's achieved
and expected rate of return. Hence, price-cap regulation is often viewed as
similar to rate-of-return regulation, however, with a longer and prespecified
regulatory lag.

Under price-cap regulation higher X and T provide higher-powered
incentives to the firm. One problem with such high-powered price-cap
incentives for electricity distribution companies lies in the inadequate
knowledge base of regulators in this industry. Another problem is the
inability of regulators to commit not to interfere when profits deviate substantially from an established norm. The difference in the commitment
problem between price caps and rate-of-return regulation is that price caps
that lead to successful innovation and investment are likely to induce the
regulator to breach the regulatory commitment by calling for an early
review or to overshoot by imposing a too stringent price cap at the next
scheduled review.

13.2.4. Profit Sharing

Under profit sharing the firm's price level is adjusted by a specified profit share times the achieved rate of return on the firm's revenues. Thus, if the firm's profit is π, its revenues are R, and the sharing parameter is s, then the firm has to reduce its prices next period on average by a factor of $s\pi/R$. In an expost view, profit sharing makes the rate payers shareholders of the company while, seen ex ante, it is a sharing of risk and an incentive device. For sufficiently short lag periods, the larger s, the smaller the incentive of the firm to reduce costs and the smaller the risk faced by the firm. With s approaching 1 the firm can keep any excess profits for only one period but also has to face losses only for one period. In this case profit sharing approaches cost-plus regulation without the plus. With s vanishing, profit-sharing approaches total deregulation of prices or pure price-cap regulation (with an infinite regulatory lag). For $0 < s < 1$ profit-sharing regulation has a number of interesting properties, assuming a firm that maximizes the discounted stream of profits:

- In a stationary environment, profits converge to 0 over time. If s is sufficiently large, the firm may engage in pure waste at the beginning of the convergence process. This holds in particular for a firm that knows in advance that this type of regulation will be introduced.[4]
- In a changing environment profit sharing has long-run dynamic effects that feed back into incentives. For example, a one-time cost reduction leads to a simultaneous profit increase that, in the next period, is followed by a price decrease with a simultaneous loss. This loss then triggers gradual price increases over time, converging to zero profits. Compare this to a permanent cost reduction, which will lead to a one-time profit increase followed by a gradual price and profit reduction over time. If several cost and demand changes occur simultaneously or in short order, there will be compound effects that can be hard to predict. Profit sharing reduces these effects over time, with the speed of reduction depending on the profit share. However, there may be some risk of long-term losses to the firm. This specifically holds for inflation, which the firm may never catch up with. That is why loss sharing may have to be excluded, or profit-sharing regulation may have to be combined with some adjustment formula for inflationary or input price changes.
- In terms of commitment problems profit sharing may outperform other types of incentive regulation because of its built-in fairness and self-correction. The regulated firm is allowed to keep only part

of its profits from windfall or superior efficiency; and consumers almost immediately share in the benefits. Consumers, therefore, can be happy about large profits because they trigger subsequent large price reductions.

- Profit sharing can have similar incentive power to price caps, depending on the length of the regulatory period and on the sharing parameter. A major difference, however, is that under price caps there is no adjustment for profit changes for T periods and then there is likely to be a large or full adjustment, while under profit sharing there is a partial adjustment in each of the T periods or a subset thereof. Under profit sharing the incentive for cost reductions that only last one period is simply reduced by the share parameter, while for permanent cost reductions the share parameter acts like an additional discount rate and thereby reduces the incentive substantially more.[5]

13.2.5. Yardstick Regulation

Under yardstick regulation the firm's price level is capped in reference to some achieved level of average cost of related firms. To the extent that these firms face similar demand and cost functions and are subject to the same random shocks, yardstick regulation allows the regulator to provide optimal incentives and leave the regulated firms no rents. The problem is that regulated firms, even in narrow sectors such as electricity transmission or distribution, can face vastly different demand and cost functions and can be subject to idiosyncratic shocks. In such a case yardstick regulation loses its effectiveness in providing incentives and limiting firm rents. However, there may remain areas for yardstick regulation to be applied as a partial measure, for example, for input price changes.

13.2.6. Discussion of Price-Level Regulation

Our discussion of types of price-level regulation reveals that none of them is ideal but that different mechanisms succeed or fail under different circumstances. The question is whether hybrid schemes can make up for deficiencies by building on the strengths of individual mechanisms and avoiding their weaknesses.

What runs under the name of price-cap regulation, as it is practiced in several countries, is already such a hybrid in that it combines aspects of rate-of-return regulation, profit sharing, yardstick regulation, and pure price caps. For electricity transmission and distribution companies we suggest such a hybrid scheme with profit sharing as its main component.

The most difficult incentive issue for distribution companies is the purchasing decision of electricity generation and transmission. With its purchasing decisions the distribution company has to set the right incentives for investments by generation and transmission companies. Thus, simply passing through the cost of electricity will provide too little incentive, although some regulatory lag, as under rate-of-return regulation, may help. On the other hand, capturing the cost of electricity in the inflation-cum-productivity adjustment of a price-cap formula would be too risky for the distribution companies, since they are not doing generation and transmission themselves. A profit-sharing mechanism is likely to represent a middle ground here in that it combines some incentive with some risk sharing. The less a country can commit to a regulatory scheme, the larger the consumer share of profits that would be optimal. In addition, with high administrative capacity a regulatory agency could offer menus of regulatory options from which the electricity distribution company could choose.

Since transmission companies offer only transmission of electricity and do not act as wholesalers of electricity, they do not face the same kind of procurement problems as distribution companies. Transmission companies, however, through their investment and pricing have some influence over the amount of electricity transmitted through their network. For given investment their total costs are largely sunk, making capacity utilization their major short-run problem. Their major long-run problem is optimal investment, optimizing over the amount of expansion and minimizing costs of investment. Risk sharing in this activity again makes profit-sharing regulation look like a good compromise.

Regulation would begin with a determination of starting prices, usually the prices ruling before the introduction of the new regulatory regime. Subsequent changes in price levels could be predetermined by a price-cap formula adjusting for input price changes and a productivity commitment. Profit sharing would come in at prespecified dates of regulatory review. The amount of productivity commitment could be traded off against consumers' profit share in a menu offered to the regulated firm. Higher productivity (or purchased-power price) commitments would be associated with lower consumer shares in profits. Where menus are administratively infeasible, the productivity commitment should probably be set at $X = 0$, and profit sharing should happen on an annual basis plus some input cost adjustment.

13.3. Price-Structure Regulation

Disputes about the type of price-structure regulation arise with respect to consumer groups and a regulated firm's competitors. Consumer groups generally want the regulated firm to charge low prices for them relative to others, while the regulated firm's competitors want the firm to charge high prices in the markets where they compete. Under monopoly regulation there are no such competitors. However, some consumers may want to bypass the distribution network, and some distribution companies or large industrial consumers may want to bypass the public transmission network. These potential bypassers may compete with other firms that cannot bypass the network. In this case they may want the regulated firm to charge high prices so that they can bypass the network whereas their competitors cannot. The bypass issue shows that the regulation of price structure creates its own kind of commitment problem. If the price structure is heavily (cross-) subsidized, incentives for entry or bypass may be created or enhanced, thus stranding investments by regulated firms.

The four types of regulating the monopolist's price level can be combined with several different types of regulating its price structure. We briefly consider six main types: cost-attribution formulas, completely frozen price structures, complete flexibility in the price structure, numerically pre-specified bands, economically specified bands, and flexibility for optional pricing.

13.3.1. Cost-Attribution Formulas

Cost-attribution formulas determine price structures by distributing the costs of the firm among its outputs and then make price structures depend on the costs thus allocated. This is usually known as fully distributed cost pricing. There are many ways to distribute not directly assignable costs among outputs, which makes this procedure very arbitrary. However, it does have a long tradition in accounting and regulation and therefore is often the status quo against which new suggestions have to be measured. In particular, rate-of-return regulation in the United States has traditionally been associated with fully distributed cost pricing.

13.3.2. Completely Frozen Price Structures

Completely frozen price structures are sometimes politically desirable because they appear to distribute price increases equally. In terms of efficiency

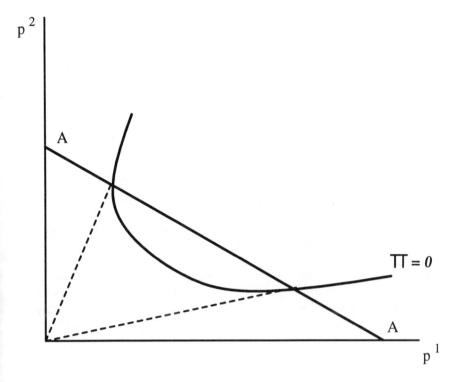

Figure 13.1.   The Flexibility of Price Structure

they make sense only if the price structure is optimal to begin with and if there are no changes in demand and cost structures. Since the starting point is usually a fully distributed cost pricing structure and since cost and demand change over time, this is usually not a desirable option.

13.3.3. Complete Flexibility

Complete flexibility in the price structure, in contrast, sounds like lack of regulation. It is, however, in practice severely limited by price-level regulation in conjunction with the firm's costs and demands (recall that competition is not an issue).

This is seen in Figure 13.1, which shows prices of two outputs of a regulated firm. The slope of a ray through the origin gives the price structure. In Figure 13.1, the price level is a price index, represented by the straight, downward sloping line AA. The slope of this line gives the relative weights of the two outputs in the index. Now add isoprofit contours, like $\pi = 0$. If the firm has to stay on or below the price level AA and at least wants to break even it is limited to price structures between the two dotted lines through the origin. This choice obviously becomes more constrained the tighter the constraint on the price level.

An unconstrained profit-maximizing monopoly firm will implement an efficient price structure, though at an inefficiently high level. By providing the right constraint on the firm's price level, the regulator can benefit from the firm's natural tendency toward an optimal price structure. For example, the V-F mechanism (Vogelsang and Finsinger, 1979) is a simple regulatory scheme that, over time and in an unchanging environment, converges to Ramsey prices. Under the (multiproduct) scheme the firm is allowed to set any prices in period t, that would result in no excess accounting profit if applied to output cost and quantities of period $t - 1$. This means that the firms face a price cap defined by a Laspeyres price index on their outputs where the cap moves from period to period by a percentage equal to the firm's previous overall profit margin. Under such a Laspeyres price index as a constraint on its output price level, the firm can achieve improvements of consumer welfare for any change in the firm's price structure, as long as consumer demands do not change. To see this, assume that a Laspeyres price index is kept constant and the price structure is changed. In order to keep the Laspeyres index constant the firm has to change prices in such a way that the consumers can still buy the old output bundle at the same total outlay and be as well off as before. Hence, by choosing a different output bundle at the new price structure the consumers must be better off in the aggregate.

This is the principle used under the V-F mechanism, and it can be achieved with profit sharing as well. However, fairness concerns or other than efficiency concerns may impose additional constraints on the firm's flexibility to choose its price structure. Also, changes in the firm's cost structure and in demand curves can occur at faster rates than changes in price structure triggered by a Laspeyres price index constraint (see, for example, Neu, 1993).

13.3.4. Numerically Prespecified Bands

Numerically prespecified bands allow the firm some limited flexibility in changing its price structure while giving consumers assurances that they are protected from large price increases. In contrast to complete flexibility in the price structure, the regulator can commit more easily to prespecified bands. In the case of electricity transmission or distribution companies lower limits are probably unnecessary because predatory pricing is of no concern.

13.3.5. Economically Specified Bands

Economically specified bands are defined by economically meaningful upper and lower limits on prices. Baumol and Willig, for example, have argued for stand-alone costs as an upper limit and incremental costs as a lower limit. The economic argument here can be made in terms of cross-subsidization and competition. Prices above stand-alone costs subsidize others and could never be maintained indefinitely with free entry. Prices below incremental costs are subsidized by others and would never be maintained indefinitely by a profit-maximizing firm (and could not be maintained by a multiproduct firm in contestable markets). Although these upper and lower bounds are economically compelling and appear to have fairness acceptance, they suffer from measuring problems for both incremental and stand-alone costs. In electricity transmission and distribution these problems are not yet solved on a practical level (see, for example, Einhorn, 1990; Hogan, 1992; Joskow, 1993). This approach, therefore, appears to be more appropriate for addressing antitrust concerns, which are treated on a case-by-case basis, rather than for ongoing monopoly regulation.

13.3.6. Optional Prices

Flexibility for optional prices allows the firm to offer consumers (nonlinear) price options in addition to regulated prices. Thus, there exists a (possibly frozen) regulated price structure along with an unregulated optional price structure. Since each customer has the option always to buy at the regulated prices, customers are protected. At the same time the firm can increase its sales and its customer base through attractive offerings. This could become important for an electricity distribution company faced with bypass.[6]

In some sense optional prices in combination with mandated prices are the dominant choice for a regulated price structure because they represent a Pareto improvement over the mandated prices. However, the regulation of a mandated (linear) price structure needs to be chosen carefully. We have already argued against a completely rigid price structure, fully distributed cost pricing, and incremental or stand-alone cost bands. This leaves us with the choice between complete flexibility and fixed upper bands. To the extent that we want to obey a fairness/equity postulate, we choose the fixed upper bands with downside flexibility and combine this with optional prices. Thus, the firm has to offer a set of tariffs fulfilling the banding requirements, and everybody will be able to buy at these tariffs. In addition, the firm may offer optional tariffs that the buyers can choose instead of the mandatory rates. The mandated prices would ordinarily protect small or average buyers, while the optional rates benefit large consumers and enhance the regulated firm's profits.

13.4. Combining Price-Level and Price-Structure Regulation

Essentially, any type of price-level regulation can be combined with any type of price-structure regulation. Hence, based on combinations of these ideal types alone we can differentiate between many types of price regulation. However, in our discussion we treat price level regulation and price structure regulation as two separate tasks. We do this in analogy to what Laffont and Tirole (1990) have called the *incentive-pricing dichotomy*. According to the incentive-pricing dichotomy, under certain conditions, incentive and pricing problems can be separated.[7] In the Laffont and Tirole framework incentives are then provided through subsidies (a cost-reimbursement rule), while price markups over marginal costs follow a symmetric-information approach. Since we do not allow for external subsidies, our regulator has to provide the optimal cost-reducing incentives via the firm's price level, then choose the best symmetric-information price structure and then combine the two.

We have chosen to combine two blended schemes, where price-level regulation is based on profit sharing, combined with features of price caps, while price structure regulation is based on setting of mandated rates with some individual caps and the possibility to use optional prices. For distribution companies profit sharing will mainly benefit consumers by passing on productivity increases and benefits from well-crafted electricity supply contracts. Under profit sharing with flexibility in the price structure the transmission companies will receive incentives to invest in transmission capacity that minimizes the combined costs of generation and transmission. At the same time, the risk faced by distribution or transmission companies is contained under profit-sharing in that consumers bear the full losses with a lag. In addition, profits derived from changes in price structure and from optional pricing will ultimately revert to consumers.

Thus, the proposed combination of regulatory instruments should balance well the incentive and risk-sharing aspects with a protection of consumers from monopolistic exploitation.

Notes

1. For other economic properties of electricity distribution companies, see Einhorn (1993).

2. In the United Kingdom the twelve distribution companies jointly own the national transmission grid through a holding company. Transmission and distribution are independently regulated.

3. In fact, we may not even consider this to be a constraint on the incentive schemes available but rather to be an implication of the fairness postulate in the objective function.

4. Proof: In a stationary environment the profit share is a multiplier similar to a discount factor. Therefore profit sharing is exactly the same as the Vogelsang and Finsinger mechanism, but for a smaller effective discount factor. As a consequence, the results of Vogelsang and Finsinger (1979) and Sappington (1980) carry over. For a short description of the Vogelsang and Finsinger mechanism, see below.

5. Let us consider the cost-reducing incentive properties of profit sharing versus price caps. Assume that price caps have a fixed regulatory lag of T periods after which prices are adjusted to produce zero profits (at period T average cost). Also assume that the firm faces completely inelastic demand so that price changes directly translate into proportional revenue changes. Now assume that the firm can reduce its cost through unobservable effort e, leading to a permanent cost reduction dC. At a given price this results in a discounted profit increase $d\pi = dC/r$. Under price caps an extra amount of dC will be earned by the firm for T periods and then revenues will be adjusted by dC.

The total effect then is $d\pi = dC (1 - g^T) / (1 - g)$ where $g = 1/(1 + r)$. Under profit sharing the firm's profits after the first period would shrink at the rate s per period and the total effect would be $d\pi = dC/[1 - (1 - s) g]$. Thus, the incentives of price caps and profit sharing with respect to such a cost-reducing innovation are equal if the effect of a cost reduction on discounted profit is the same or if $s = [(1 - g)/g][g^T/(1 - g^T)] = r/[(1 + r)^T - 1]$. Keeping incentives for price caps and profit sharing at the same level we get the following tradeoffs: s is declining in T with $s = 1$ for $T = 1$ and $s = 0$ for $T =$ infinite. That means, the larger the regulatory lag under price caps, the smaller the equivalent profit share going to consumers. The effect of discounting on the equivalent profit share of consumers (for given T) is not monotonous. However, for sufficiently small discount rates it is usually decreasing. That means, the more the future is discounted the smaller the equivalent profit share going to consumers. In case of no discounting we have $s = 1/T$. If we want to provide the same cost-reducing incentives under profit sharing as under price caps and if $T = 4$ and $r = 10$ percent, then we have to set $s = 20$ percent. For the type of innovation of our example the popular 50 percent sharing rule could provide less cost-reducing incentives than a two-year price cap. However, this effect depends on the specific simplifications made for this analysis. In particular, note that the relative incentive properties of profit sharing and price caps are likely to be different for one-time cost reductions.

6. Optional pricing also affects the firm's price level. However, optional prices are nonlinear, and the theory of nonlinear price indices is not well developed. The average price

derived by dividing consumer outlays by quantities can be used as an imperfect substitute for a nonlinear price schedule when it comes to constructing price indices.

7. The main condition is that the firm's type and effort can be aggregated in the firm's cost function (see Laffont and Tirole, 1990, p. 17).

References

Baumol, W.J., J.C. Panzar, and R.D. Willig. 1982. *Contestable Markets and the Theory of Industry Structure*. New York: Harcourt Brace Jovanovich.

Brown L., M.A. Einhorn, and I. Vogelsang. 1991. "Toward Improved and Practical Incentive Regulation." *Journal of Regulatory Economics*, 3: 313–338.

Crew, M.R., and P. Kleindorfer. 1992. "Incentive Regulation, Capital Recovery and Technological Change in Public Utilities." In M.A. Crew (ed.), *Economic Innovations in Public Utility Regulation* (pp. 57–79). Norwell, Mass.: Kluwer.

Einhorn, M.A. 1990. "Electricity Wheeling and Incentive Regulation." *Journal of Regulatory Economics*, 2: 173–189.

———. 1993. "Incentive Regulation for Power Distributors in the British Electricity Experiment." Mimeo, U.S. Department of Justice, Washington, DC.

Gilbert, R., and D. Newberg. 1988. "Regulation Games." Discussion Paper No. 267, Centre for Economic Policy Research, London (September).

Hogan, W.W. 1992. "Contract Networks for Electric Power Transmission." *Journal of Regulatory Economics*, 4: 211–242.

Joskow, P.L. 1974. "Inflation and Environmental Concern: Structural Change in the Process of Public Utility Price Regulation." *Journal of Law and Economics*, 17: 291–328.

———. 1993. "Electricity Agenda Items for the New FERC." *Electricity Journal* (June): 18–28.

Joskow, P.L., and P. MacAvoy. 1975. "Regulation and Franchise Conditions of the Electric Power Companies in the 1970s." *American Economic Review*, 65: 295–311.

Joskow, P.L., and R. Schmalensee. 1983. *Markets for Power*. Cambridge, MA: MIT Press.

Laffont, J.J., and J. Tirole. 1990. "The Regulation of Multiproduct Firms. Part I: Theory." *Journal of Public Economics*, 43: 1–36.

Levy, B., and P. Spiller. 1993. "Regulation, Institutions, and Commitment in Telecommunications: A Comparative Analysis of Five Country Studies." Paper Prepared for the World Bank's Annual Conference of Development Studies, Washington, D.C., Page 304.

Lyon, T.P. 1991. "Regulation with 20-20 Hindsight: 'Heads I Win, Tails You Lose.'" *Rand Journal of Economics*, 22: 581–595.

Neu, W. 1993. "Allocative Inefficiency Properties of Price-Cap Regulation." *Journal of Regulatory Economics*, 5(2) (June): 159–182.

Vogelsang, I., and J. Finsinger. 1979. "A Regulatory Adjustment Process for Optimal Pricing by Multiproduct Monopoly Firms." *Bell Journal of Economics*, 10: 157–171.

Author Index

Subject Index